AN UNOFFICIAL HISTORY OF
NASA MISSION PATCHES

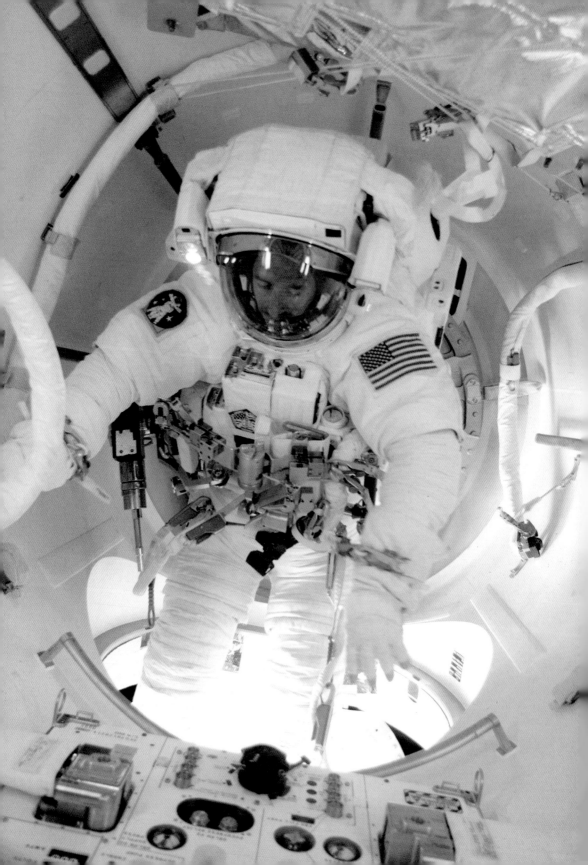

AN UNOFFICIAL HISTORY OF
NASA MISSION PATCHES

ROGER D. LAUNIUS

San Diego, California

Thunder Bay Press
An imprint of Printers Row Publishing Group
10350 Barnes Canyon Road, Suite 100, San Diego, CA 92121
www.thunderbaybooks.com • mail@thunderbaybooks.com

Copyright © 2020 Quarto Publishing plc

All rights reserved. No part of this publication may be reproduced, distributed, or transmitted in any form or by any means, including photocopying, recording, or other electronic or mechanical methods, without the prior written permission of the publisher, except in the case of brief quotations embodied in critical reviews and certain other noncommercial uses permitted by copyright law.

Printers Row Publishing Group is a division of Readerlink Distribution Services, LLC. Thunder Bay Press is a registered trademark of Readerlink Distribution Services, LLC.

Correspondence regarding the content of this book should be sent to Thunder Bay Press, Editorial Department, at the above address. Author, illustration, and rights inquiries should be addressed to The Bright Press at the address below.

This book was conceived, designed, and produced by
The Bright Press, an imprint of The Quarto Group
The Old Brewery, 6 Blundell Street, London, N7 9BH
www.quartoknows.com

Thunder Bay Press
Publisher: Peter Norton
Associate Publisher: Ana Parker
Acquisitions Editor: Kathryn Chipinka Dalby
Editor: Dan Mansfield

The Bright Press
Publisher: James Evans
Editorial Director: Isheeta Mustafi
Managing Editor: Jacqui Sayers
Senior Editor: Caroline Elliker
Art Director: Katherine Radcliffe
Book design and layout: gradedesign.com
Cover design: Greg Stalley

The trademarks, names, and logos on the patches featured in this book are the property of NASA. This book is not an official publication of NASA and it has not been prepared, approved, endorsed, or licensed by NASA.

ISBN: 978-1-64517-415-8

Printed in China

24 23 22 21 20   1 2 3 4 5

CONTENTS

FOREWORD	6	CHAPTER 7 SPACE SHUTTLE 1994-1998 »	116
INTRODUCTION	8		
CHAPTER 1 PROJECT MERCURY »	18	CHAPTER 8 SPACE SHUTTLE 1999-2011 »	132
CHAPTER 2 PROJECT GEMINI »	28	CHAPTER 9 ISS EXPEDITIONS 2000-2010 »	158
CHAPTER 3 PROJECT APOLLO »	42		
CHAPTER 4 PROJECT SKYLAB and ASTP »	68	CHAPTER 10 ISS EXPEDITIONS 2011-2020 »	178
		CONCLUSION	200
CHAPTER 5 SPACE SHUTTLE 1981-1986 »	78	FURTHER READING GLOSSARY OF TERMS PICTURE CREDITS	202 204 205
CHAPTER 6 SPACE SHUTTLE 1988-1993 »	98	INDEX ACKNOWLEDGMENTS	206 208

FOREWORD

I wore a space shuttle mission patch for the first time in 1981. It felt really good to have that emblem on my flight suit, even though my name was not on it and I wasn't going on the flight. That STS-2 patch signified that I was part of the mission support team, in my case, assigned to fly in the chase plane that would accompany *Columbia* through its final approach to landing and capture data about the fragile thermal protection tiles on the shuttle's belly. Wearing a crew patch was a great honor, one that was reserved—at least within NASA—to the flight crew, their families, and the much larger team of engineers, technicians, and scientists involved directly in mission preparation and flight operations. I wore five other patches without my name on them over the next two years, as I served in support roles for the third through seventh shuttle missions. Finally, in late 1983, I sat down with my STS-41G crewmates to begin designing the emblem that would signify our mission, the first one that would bear my name. It was an unforgettable moment.

The shuttle program almost didn't have individual mission patches. The shuttle was supposed to make spaceflight as routine as airline operations and be focused on the customers, rather than the astronauts. The "Right Stuff" era was over, and NASA higher-ups felt the time had come to do away with the personalization of mission patches. Shuttle crews would wear the generic triangular logo of the space shuttle program instead, perhaps adding a small rectangle below it that bore the mission number. That idea went over like a lead balloon in the astronaut corps, with the results shown in Chapter 6 of this volume.

That was not the first time there was a tussle about technical versus personalized designations of space missions, as Roger Launius's narrative makes clear. Behind every emblem in this delightful book is a story rich in symbolism and emotion. Some of these were particular to a specific astronaut. Others, most notably the Apollo 11 patch, spoke to a global audience. All held great meaning to the many hundred people behind each mission, and continue to inspire us today. I hope you enjoy the walk down memory lane as much as I did.

Kathryn D. Sullivan

1]
Kathryn Sullivan during her historic space walk in October 1984. Here she checks the latch of the SIR-B antenna in *Challenger*'s open cargo bay.

2]
Kathryn Sullivan uses a pair of binoculars to look through the forward cabin windows of the space shuttle.

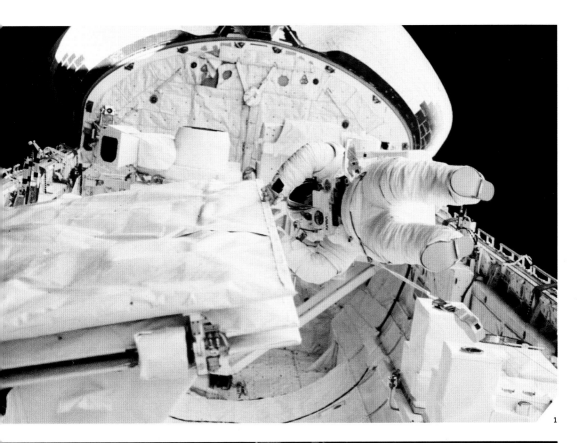

HUMAN SPACEFLIGHT

Mission patches are ubiquitous in NASA history. They have defined the human spaceflight effort since the 1960s. They epitomize the first decade of the Space Age in the United States just as much as the aggressive human space missions they celebrated: Mercury's single astronaut program (six piloted flights during 1961–1963) to ascertain if a human could survive in space; Project Gemini (10 two-astronaut flights during 1965–1966) to practice rendezvous and docking of spacecraft and extravehicular activity (EVA); and Project Apollo (10 three-astronaut flights during 1968–1972) to explore the Moon.

Project Apollo resulted from circumstances that were unique to the spring of 1961 and occurred at no other time during that decade. The result was President John F. Kennedy standing before Congress and the nation on the afternoon of May 25, 1961, announcing, "I believe this nation should commit itself to achieving the goal, before this decade is out, of landing a man on the Moon and returning him safely to the Earth." It was a bold announcement, but one he made tentatively. In the end NASA expended $25.4 billion in the 1960s—about $180 billion in 2020 dollars—and the majority of the American public believed it was too much. There were attacks on the NASA budget every year from 1962 through the end of the Apollo program. I believe it would have been a bargain at twice that price.

This was largely because of the Cold War challenges it was designed to overcome. In the early 1960s the world was locked in a superpower struggle between two nations with diametrically opposed political and economic ideologies. Like two scorpions in a bottle, either the Soviet Union or the U.S. was going to win that Cold War. Project Apollo helped gather international allies to the Western coalition that opposed the Soviet Union. Everyone recognized then, as many still do today, that the people who can master science and technology are the people that will lead us into the future. There was no greater demonstration of science and technology than the Moon landing. It impressed nonaligned nations enough for them to throw in their lots with the U.S. alliance. In no small measure, it helped to win the Cold War.

1]
The launch of the first orbital mission of Project Mercury on February 20, 1962, when astronaut John Glenn orbited the Earth.

2]
Astronaut John Young, commander of the *Apollo 16* lunar landing mission in 1972, jumps up from the lunar surface to salute the U.S. flag.

3]
Astronaut Edward White's *Gemini IV* space walk on June 3, 1965.

AFTER APOLLO

NASA clearly had high hopes of a terrific encore after the ending of the Apollo program in December 1972. It had advocated for the building of winged, reusable spacecraft that would make going to and from space relatively easy. It had designs on a lunar base to be established in the next decade, and a Mars mission before the end of the 20th century. The best it could do was undertake the *Skylab* orbital workshop program with repurposed Apollo hardware, a diplomatic mission to rendezvous and dock with a Soviet spacecraft, and to build a partially reusable space shuttle. The space shuttle emerged in the 1970s as the premier human space activity for the U.S., and when it began flying in 1981, it captured the attention of a new generation watching its astronauts in orbit.

The first flight of *Columbia* on April 12, 1981, was a rousing event. It was twenty years to the day since Yuri Gagarin's first orbit of the Earth. Walter Cronkite, covering the event for CBS News, could barely contain himself at the launch as he watched an entirely new type of space vehicle—one with wings and wheels that could be reused many times—take off for space. He showed even more excitement two days later as *Columbia* returned to Earth, landing like an aircraft at Edwards Air Force Base in the high desert of Southern California. He gushed, and he was not alone; the new age of human space exploration was upon us.

NASA shuttle astronauts were much more than the tough-as-nails, ex-fighter pilot stereotypes. Some, like John Young, had an impressive pedigree that extended back to the Gemini and Apollo programs of the 1960s. The public immediately took a liking to those fliers. But the shuttle program also allowed a new diversity in the astronaut corps that had never been seen before. The inclusion of women and people of color in the astronaut corps meant that it now, more than ever before, looked like the rest of America. But NASA recruited a diversely skilled set of astronauts as well, especially scientists of all types. STEM (science, technology, engineering, and math) professionals could now aspire to become astronauts and explore the cosmos. And they did. Over 30 years of space shuttle operations, hundreds of astronauts have flown, and their scientific work has changed human understanding of the universe. In 2011, the last of 135 space shuttle missions signaled the end of the longest and most multifaceted human spaceflight program in American history.

4]
Columbia is being transported to the launch complex for the launch of STS-1. The cars and trucks on the ground give a sense of scale.

5]
The space shuttle *Endeavour* is photographed performing a backflip to inspect the shuttle's heat shield during the STS-127 mission on July 17, 2009.

6]
Astronaut Owen Garriott performs a space walk to retrieve film from the Apollo Telescope Mount outside the *Skylab* orbital workshop in 1973.

4

5

6

AT HOME IN SPACE

INTRODUCTION

In 1984, as part of its interest in reinvigorating the space program, the Reagan administration called for the development of a permanently occupied space station. Initially projected to cost $8 billion, within five years the projected costs had more than tripled and the station had become too expensive. NASA pared away at the station budget, and soon the project was satisfactory to almost no one. In 1993, the end of the Cold War allowed NASA to negotiate a landmark decision to include Russia in the building of the International Space Station (ISS). By 1998 the first elements had been launched and in 2000 the first crew went aboard. By the beginning of the twenty-first century, the effort involved fifteen nations, and despite the end of the space shuttle program in 2011, concerted efforts took place to complete the ISS.

The *Columbia* accident of 2003, which resulted in the deaths of seven astronauts, grounded the space shuttle fleet and thereby placed on hold the construction of the ISS. Access to the station, thereafter, came only through the use of the Russian Soyuz capsule, a reliable but limited vehicle whose technology extended back to the 1960s. Because of this limitation, the ISS crew was cut to two members in May 2003—a skeleton workforce designed to keep the station operational.

When the space shuttle resumed operations in 2006, efforts to complete the ISS intensified. Throughout the ISS program, a series of crews—called Expeditions—occupied the ISS. It has become the orbital home to world-class scientific research, especially materials science and the human response to microgravity.

These efforts continue to the present, and before the ISS reaches the end of its service life in 2030, it may well revolutionize scientific understanding.

7] Against the backdrop of the Earth, an experiment on the exterior of the ISS collects information on how different materials react in space in August 2007.

8] In December 1998, the crew of STS-88 began construction of the ISS, joining the U.S.-built Unity node to the Russian-built Zarya module.

9] During the flight of STS-130, NASA astronaut George Zamka is pictured in a window of the newly installed Cupola on the ISS.

7

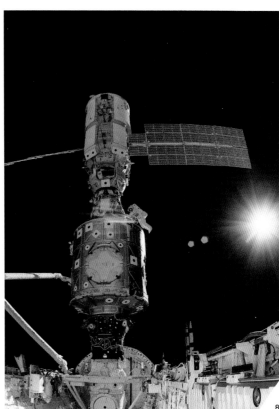

8

9

MEMENTOS IN TIME

INTRODUCTION

Perhaps the most readily identified symbol of every space mission is the mission patch. It has become omnipresent in the history of the U.S. space program. Mission patches adorn the astronauts' space suits, flight suits, clothes, mission briefing packets, publications, and a host of other items. These may be purchased in every NASA visitor center, air and space museums around the world, and from many websites. They help to personalize the story of human space exploration, tying it closely to the astronauts and the objective of individual spaceflights.

The first mission patches were actually created and worn by Soviet cosmonauts in 1963. The first ever mission patch was worn by Soviet cosmonaut Valentina Tereshkova on the *Vostok-6* mission in 1963. American astronauts, most of whom were military pilots, took their tradition of shoulder patches for military units with them into space.

Beginning with the Gemini program of the mid-1960s, the astronauts began to create mission patches for each flight. As astronaut Stephen Robinson, STS-114 mission specialist, commented: "The patch is a big deal to every crew. It's probably the most visibly unique symbol of the mission, usually tells a story about the mission and it is often a peek into the personalities of the people onboard." Robinson had a major role in the design of the STS-114 mission patch, the first space shuttle flight after the tragic accident involving *Columbia* in 2003.

NASA astronauts consider it a major aspect of their mission preparation to design a patch. There is no standard approach. Often they formulate as a group the major ideas they want to convey in the patch, and then assign one or more members of the crew to work with a designer to realize the vison. Usually the designer is a staff artist for NASA, but some famous artists from outside the agency have also created patch designs.

The design is an iterative process, with the designer taking direction from the crew. It also has to be approved by NASA's management. This is usually pro forma, but occasionally the crew has to make further modifications. Since the 1960s, mission patches have come to define spaceflight, serving as mementos in time for both the astronauts and the public.

10]
(From left to right): Astronauts Fred Haise, Jack Swigert, and James Lovell pose with the *Apollo 13* patch and spacecraft replicas before launch in April 1970.

11]
Renowned artist Robert McCall works on a mural of STS-1 that includes the mission patch he designed. McCall spent more than 50 years creating powerful visual imagery of space exploration.

COLLECTING PATCHES

Collectors avidly acquire mission patches; many are readily available but some are rare. Several manufacturers make versions of these patches, and seeking out the various versions, learning about their differences, and expanding the cache of knowledge about them motivate collectors. Subtle differences are largely unseen by the casual observer, but collectors have drawn up detailed explanations of each version of every individual mission patch.

Two companies have been especially significant in the manufacturing of mission patches, and the differences between them factor prominently in their attractiveness for the collector. A-B Emblem of Weaverville, North Carolina, has made most of the official versions used by the astronauts. Lion Brothers of Owings Mills, Maryland, however, also makes many patches and agency contractors have often procured patches from them. A standing contract between NASA and A-B Emblem dates from February 1970 for its human spaceflight mission patches.

Whether or not a mission patch is official helps to determine its significance as a collectible item, but the definition of what is or isn't official is quite murky. Embroidered patches were often procured by NASA and the major contractors independently of each other. Different suppliers were used, even for the same mission. Some collectors argue that patches given out by NASA and contractors are, by definition, official; other items available for public sale are unofficial. Subtle differences between patches, the rarity of patches with these divergences, and the provenance of the item all affect the desirability of the collectible.

Because of the risk of fire on a space mission, NASA had Beta cloth versions of mission patches manufactured by Owens-Corning Fiberglass of Ashton, Rhode Island, under contract to NASA. The patch designs were silk-screened (using hand-mixed pigments from Roma Color of Fall River, Massachusetts) onto Beta cloth. Some of those are in the collector's market, and are highly prized as space-flown items.

In addition to being worn by crew, these mission patches are usually worn on garments of ground support staff and contractor personnel to signify playing an active role in support of a mission. These are also desired memorabilia. For those with an interest in collecting, there are several publications and websites with information of use. See Further Reading (page 202).

12]
Apollo 16 crew members John Young, Ken Mattingly, and Charlie Duke walk out on launch day, April 16, 1972. The mission patch is prominent on the van waiting to take them to the launch complex.

13]
The European Space Agency's mission patches for the astronaut class of 2009 were placed on the ISS observation window by ESA astronaut Thomas Pesquet.

After several delays and more than four hours in the *Freedom 7* Mercury capsule, astronaut Alan Shepard was getting restless to become the first American in space. He told Mission Control to "fix your little problem and light this candle."

CHAPTER 1
PROJECT MERCURY >>

PROJECT MERCURY

On October 1, 1958, the National Aeronautics and Space Administration (NASA) officially began operations. It had been tasked by President Dwight D. Eisenhower with getting an American into orbit. Project Mercury was born. And over the next five years the program carried out three key phases:

1. Developing systems and technologies, as well as choosing and training astronauts
2. Using Redstone rockets to send humans on suborbital, ballistic flights
3. Launching astronauts into orbit using Atlas rockets

In addition, NASA orbited a human-piloted spacecraft around Earth, investigated a pilot's ability to function in space, and recovered both the pilot and spacecraft safely. In all, Project Mercury made six piloted flights.

The program was named after the Roman god of trade, merchants, thieves, travelers, and communication by Abe Silverstein, director of the NASA Lewis Research Center in Cleveland. Silverstein recommended the allegorical and evocative name, thinking that Mercury, the son of Jupiter and grandson of Atlas, whose winged sandals and helmet reinforced the idea of speed, was too symbolic a name not to use. Accordingly, on the anniversary of the first flight of the Wright brothers, December 17, 1958, NASA announced its first human spaceflight program would be called Project Mercury.

1]
The Mercury spacecraft was a Cold War expediency replacing the long-standing desire to fly a spaceplane beyond Earth going back to the science-fiction stories featuring Buck Rogers and Flash Gordon. Because capsules were easier to build than spaceplanes, the small, conical capsule safeguarded the first Americans who went into space. This artist's concept depicting the names of each capsule and the signature of the astronaut who flew it was released by NASA in 1963.

2]
The Mercury Seven astronauts in their iconic silver space suits, 1959. From left to right, back row: Alan Shepard, Virgil "Gus" Grissom, L. Gordon Cooper; front row: Walter "Wally" Schirra, Donald "Deke" Slayton, John Glenn, and M. Scott Carpenter.

3]
Astronaut John Glenn enters his *Friendship 7* Mercury capsule to begin his historic flight as the first American to orbit the Earth, on February 20, 1962. Shown clearly below the hatch is the Friendship 7 logo designed for the mission.

MERCURY SPACECRAFT
Designed and built by MCDONNELL, St. Louis
for the National Aeronautics and Space Administration

1

2

3

Although there had been a long tradition of patches worn on military flight suits, the Mercury astronauts chose not to adopt the practice, despite all these astronauts having been recruited from the military. The original Mercury Seven astronauts wore only the NASA logo on their silver space suits. The astronauts did, however, give names to their spacecraft. Alan Shepard began the tradition with *Freedom 7*, and the other astronauts adopted the "7" in honor of the number of astronauts in the program. As Shepard commented, "Pilots have always named their planes. It's a tradition. It never occurred to me not to name my capsule. I checked with Dr. [Robert] Gilruth [head of the Space Task Group leading Project Mercury] and I talked it over with my wife and with John Glenn, who was my backup pilot. We all liked it."

Not until the Gemini program did mission patches become the norm. Since no mission patches were produced at the time of the Mercury missions themselves, any patches you see today for these missions are generally regarded as modern souvenirs. A-B Emblem, which began partnering with NASA in 1960, created the souvenir Mercury patches to fill the void since there were no official mission patch designs. There are sets packaged with TWA Tours labeling, which would date them to the earliest Kennedy Space Center gift shops of the mid-1960s. By the time of the Apollo missions, during the late 1960s, several companies began to produce unofficial Project Mercury commemorative patch sets for public sale.

4

5

4]
This NASA insignia was the only patch worn on the space suits of the Mercury astronauts.

5]
The first astronauts, universally called the Mercury Seven, helped originate this design for the Mercury program in 1964. The symbol represented around the 7 is the astronomical symbol for the planet Mercury, and the medieval sign for the element mercury as well. It consists of the biological sign for female, topped with "horns" to represent the winged hat that Mercury is usually depicted as wearing.

Opposite]
Mercury astronauts (from left to right) John Glenn, Gus Grissom, and Alan Shepard standing by the Redstone rocket in their space suits in 1961.

MERCURY-REDSTONE 3

LAUNCH DATE May 5, 1961
LAUNCH VEHICLE *Freedom 7*
FLIGHT TIME 15 minutes
CREW Alan Shepard

With 45 million Americans watching on TV, Alan Shepard's flight came less than three weeks after Soviet cosmonaut Yuri Gagarin became the first person to reach outer space. *Freedom 7* was a ballistic "cannon shot" up and down. Shepard traveled 116.5 miles high and 302 miles downrange from Cape Canaveral. Comparisons between the Soviet and American flights inevitably followed. Gagarin had orbited the Earth; Shepard did not. Gagarin's *Vostok* spacecraft weighed just 428 pounds; *Freedom 7* weighed 2,100 pounds; Gagarin had been weightless for 89 minutes; Shepard for only 5 minutes. *Freedom 7*'s name fittingly played to the Cold War tensions between the U.S. and the Soviet Union, pitting the "free" world against the communist one.

MERCURY-REDSTONE 4

LAUNCH DATE July 21, 1961
LAUNCH VEHICLE Liberty Bell 7
FLIGHT TIME 16 minutes
CREW Virgil "Gus" Grissom

Gus Grissom's mission was largely a repeat of Shepard's with a few minor improvements, such as updated hand controllers, a window, and an explosive side hatch for quick escape in case of emergency. This hatch blew prematurely after *Liberty Bell 7* parachuted into the Atlantic Ocean near the Bahamas. The capsule sank, and Grissom nearly drowned before being hoisted to safety in a helicopter. In 1999 Curt Newport led a team to recover *Liberty Bell 7* from the bottom of the Atlantic Ocean. It has now been restored and is on display at the Kansas Cosmosphere.

MERCURY-ATLAS 6

LAUNCH DATE February 20, 1962
LAUNCH VEHICLE *Friendship 7*
FLIGHT TIME 4 hours, 55 minutes
CREW John Glenn

In this first American orbital flight, John Glenn circled the Earth three times at an altitude of 162 miles. The flight was not without problems. Glenn had to fly parts of the last two orbits manually because of an autopilot failure, and he used his retrorocket pack (which would normally be jettisoned) to hold in place a potentially loose heat shield. He also ran out of fuel as he descended through the upper atmosphere. Glenn returned safely, but splashed down 40 miles short of his Atlantic Ocean target.

MERCURY-ATLAS 7

LAUNCH DATE May 24, 1962
LAUNCH VEHICLE *Aurora 7*
FLIGHT TIME 4 hours, 56 minutes
CREW M. Scott Carpenter

Scott Carpenter's *Aurora 7* mission also circled Earth three times, but it was dogged by controversy. Carpenter's spacecraft splashed down some 250 miles off-course, delaying recovery for an hour. NASA officials held Carpenter responsible for the error, accusing him of disregarding his instructions, an accusation that Carpenter vehemently denied until his death in 2013.

Carpenter had preferred "Rampart 7" for the name of his capsule after the mountain range in his native Colorado, but he decided that *Aurora 7* would "come through the static better" during radio broadcasts. "It had a sentimental meaning to me because my address as a child back in Colorado was on the corner of Aurora and Seventh Streets in Boulder," he commented.

MERCURY-ATLAS 8

LAUNCH DATE October 3, 1962
LAUNCH VEHICLE Sigma 7
FLIGHT TIME 9 hours, 13 minutes
CREW Walter "Wally" Schirra

Wally Schirra's mission focused on engineering and control. He tested the autopilot system and starfield navigation, used the first Hasselblad camera in orbit, broadcast the first live message from an American spacecraft, and made the first splashdown in the Pacific Ocean. Of his capsule's name Schirra recalled, "Sigma means 'sum of.' I wanted to get off the 'gee whiz' names and use a technical term, as well as acknowledge the original seven." He also said it helped remind everyone that the mission "was the sum of the efforts and energies of a lot of people."

MERCURY-ATLAS 9

LAUNCH DATE May 15, 1963
LAUNCH VEHICLE Faith 7
FLIGHT TIME 1 day, 10 hours, 20 minutes
CREW L. Gordon Cooper

The capstone of Project Mercury came on May 15–16, 1963, when Gordon Cooper circled Earth 22 times in 34 hours. Cooper released the first satellite from a spacecraft, a 6-inch sphere with a beacon for testing the astronaut's ability to track objects visually in space. Gordon Cooper said he chose his capsule name just a few days before the flight. "I selected the name *Faith 7*," Cooper recalled, "to show my faith in my fellow workers, my faith in all the hardware so carefully tested, my faith in myself, and my faith in God."

Gemini II is ready for launch: this was the second Titan II Gemini Launch Vehicle to set off from Earth. On board was the unmanned Gemini spacecraft for the program's first suborbital flight.

CHAPTER 2
PROJECT GEMINI >>

PROJECT GEMINI

Project Gemini originated on January 3, 1962, as a means of bridging the capability gap between what Project Mercury had demonstrated and what would be necessary to reach the Moon. The program's major objectives included:

- Rendezvous and docking
- Long-duration flights
- Extravehicular activity (EVA) and spacewalking
- Advanced, reliable on-board flight systems
- Training of flight and ground crews

The Gemini capsule weighed more than 7,900 pounds—twice the weight of Mercury—but seemed more cramped, with only 50 percent more cabin space for double the number of astronauts. Because of the more sophisticated missions of the Gemini program, the two-person capsule had a service module, with life-support equipment, expendables, and maneuvering systems.

It was during this era that NASA astronauts first designed unique mission patches to set each flight apart. This came about in part because of a controversy over the naming of spacecraft, which had been the standard for Project Mercury.

During *Gemini III*, the first flight with astronauts aboard, mission commander Gus Grissom tried to name his capsule *Molly Brown*, after the Broadway show *The Unsinkable Molly Brown*. NASA officials weren't happy with this name. They didn't want to be reminded of how Grissom's previous Mercury capsule had been lost in the depths of the Atlantic Ocean. The word came down from Houston, and after this there were no more official spacecraft names for the Gemini program.

1]
Major internal components of the Gemini spacecraft, 1966.

2]
Astronauts Charles "Pete" Conrad and Gordon Cooper aboard the recovery ship after the *Gemini V* flight. The first mission patch worn by astronauts in orbit can be seen over the right breast.

CHAPTER 2: PROJECT GEMINI

In August 1965, after the first two manned Gemini missions had flown, astronauts Gordon Cooper and Charles "Pete" Conrad requested that they be allowed to wear an individually designed patch on their flight suits for *Gemini V*. On August 16, 1965, NASA administrator James Webb approved this request, agreeing in a memo that "on GT-5 and future Gemini flights, such an identification may be worn on the right breast between the nameplate of the astronaut; said 'patch' is to be no larger than the NASA emblem worn on the left breast. This patch will be referred to by the generic name of the 'Cooper patch.' If such 'Cooper patch' is not to be worn, the designation of the flight '*Gemini VI*' or '*Gemini VII*' may be suitably put beneath the nameplate."

Thereafter, NASA's Astronaut Office contracted A-B Emblem to produce two types of mission patches for every flight. The first type was a fireproof, cloth version woven from Teflon-coated glass fibers for use on the flight suits. These cloth patches were never made available to the public. A second, souvenir embroidered patch often flew in the astronauts' personal preference kits, for distribution to friends, dignitaries, and associates. Many non-flown versions of these embroidered patches were also marketed to the public. During the years since Gemini, there has been a healthy collectors' market for mission patches. Other companies besides A-B Emblem, especially Lion Brothers, have manufactured patches for those collecting them. Whether vintage patches or recent reproductions, many of these are well-executed versions of the flown items.

3

3]
The insignia of the Gemini program shows a gold zodiac Gemini symbol. The two white stars represent the Gemini twins, Pollux and Castor.

4]
As part of their training for *Gemini VI-A*, astronauts Tom Stafford (left) and Wally Schirra (right) practice suiting up.

5]
Dr. Robert Gilruth (left), in Houston's Mission Control Center during the 1966 *Gemini XII* mission with astronauts, from right, Pete Conrad, John Glenn, and Alan Shepard.

6]
Astronauts John Young (right) and Michael Collins (left) suiting up for the *Gemini X* mission. Their mission patches are featured on their right shoulders.

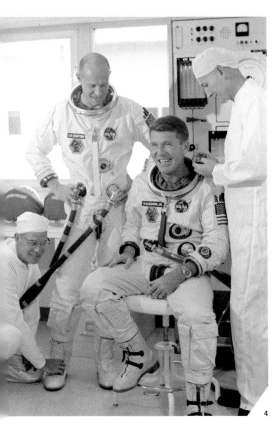

GEMINI III

LAUNCH DATE March 23, 1965
FLIGHT TIME 4 hours, 53 minutes
CREW Virgil "Gus" Grissom, John Young

This mission's primary goal involved testing the new Gemini spacecraft. In space, the crew fired thrusters to change their orbit, shift their orbital plane slightly, and drop to a lower altitude. Since reentry proved more challenging than expected, *Gemini III* splashed down 69 statute miles short of its intended landing site. The capsule was suspended at two points from a parachute, but the force of the impact cracked the faceplate of Gus Grissom's helmet.

Two different patch styles for *Gemini III* gained popularity after the flight. With the approval of mission patches for all flights beginning with *Gemini V*, designs for earlier missions were created. One showed the spacecraft in flight and the name GT-3, while the other showed the spacecraft floating in the ocean after landing, with the names of the crew around it.

GEMINI IV

LAUNCH DATE June 3, 1965
FLIGHT TIME 4 days, 1 hour, 56 minutes
CREW James "Jim" McDivitt, Edward "Ed" White

Flown by two rookie astronauts, this mission is largely remembered for the 22-minute space walk of Ed White, the first ever for an American. Tethered to the spacecraft and using a small "zip gun" to maneuver himself, White floated peacefully through space while mission commander Jim McDivitt took photographs. *Gemini IV* set a four-day, 62-orbit record for flight duration, while the new Mission Control Center in Houston oversaw its activities for the first time.

With no mission patch for this flight, the crew of *Gemini IV* wore a shoulder patch of the American flag on their suits. McDivitt recalled, "This was the first time the American flag had been worn on a pressure suit and it has continued to be used ever since. The original flags we had sewn on we purchased ourselves. Later on, of course, NASA made this an integral part of the pressure suit."

GEMINI V

LAUNCH DATE August 21, 1965
FLIGHT TIME 7 days, 22 hours, 55 minutes
CREW L. Gordon Cooper, Charles "Pete" Conrad

Gemini V set another flight duration record by more than doubling the previous time to almost eight days. Because of the longer duration planned for this mission, the crew designed a mission patch with the iconic Conestoga wagon, which they wore in memory of the overlanders of the 19th century. They persuaded NASA to allow the patch to be used on the mission, setting in motion a trend repeated by every astronaut crew thereafter.

The crew came up with the slogan "Eight Days or Bust" to be laid over the wagon, but NASA administrator James Webb insisted that they omit the slogan. "If you don't make eight days," Webb said, "I don't want the press having a field day about the mission being a bust." The crew agreed to hide the words under the canvas on the wagon, which they later removed when they were successful in the mission.

GEMINI VII

LAUNCH DATE December 4, 1965
FLIGHT TIME 13 days, 18 hours, 35 minutes
CREW Frank Borman, James "Jim" Lovell

Setting another record, this 14-day mission addressed problems of long-duration spaceflight as never before. The high point was the rendezvous with *Gemini VI-A* and flying with it for a short period. The astronauts also performed 20 experiments, the most of any Gemini mission. They tried out a new, lightweight space suit, but it proved uncomfortable, hot, and unwieldy. Astronaut Jim Lovell wrote about the mission patch that, since it was such a long mission, "We wanted an insignia that would signify medicine and endurance, much like a long-distance runner. We chose the torch as that emblem."

GEMINI VI-A

LAUNCH DATE December 15, 1965
FLIGHT TIME 1 day, 1 hour, 51 minutes
CREW Walter "Wally" Schirra, Thomas Stafford

This mission rendezvoused within 12 inches of *Gemini VII*, remaining in proximity for five hours and achieving one of Gemini's primary goals, orbital rendezvous. Wally Schirra said he designed the patch "to locate in the sixth hour of celestial right ascension. This was the predicted celestial area where the rendezvous should occur in the constellation Orion." The *Gemini VI-A* spacecraft is shown superimposed on the "twin" stars Castor and Pollux. The crew recalled, "We were up there aiming for the rendezvous and when we first saw our rendezvous vehicle, *Gemini VII*, glittering in the reflected light of the sunset, it was right between Sirius and the twins, just exactly where we had placed it on the patch."

GEMINI VIII

LAUNCH DATE March 16, 1966
FLIGHT TIME 10 hours, 41 minutes
CREW Neil Armstrong, David Scott

This mission narrowly missed tragedy when the crew attempted to rendezvous and dock with an Agena target vehicle. Although the docking went smoothly and the two vehicles orbited together, they began to pitch and roll wildly because of a stuck thruster. Armstrong undocked the *Gemini VIII* and used reentry control thrusters to regain control of his craft, but the astronauts had to make an emergency landing in the Pacific Ocean only 10 hours after launch.

The mission patch depicts light emanating from the twin stars Castor and Pollux split into a spectrum, which then divides into the symbol for Gemini followed by the Roman numeral "VIII." Neil Armstrong said that the prism "indicates that the flight objectives cover the complete spectrum of the Gemini program objectives."

GEMINI IX

LAUNCH DATE June 3, 1966
FLIGHT TIME 3 days, 21 minutes
CREW Thomas Stafford, Eugene "Gene" Cernan

By the time that *Gemini IX* flew, the Gemini program was moving on toward completion, and NASA leaders were hopeful that by the time of the last flight later in 1966 the agency would gain the experience necessary to move on to the Moon landing effort. They were taken aback, however, by the failure of the extravehicular activity by astronaut Gene Cernan on this mission. He was to leave the spacecraft, make his way to the back of the service module, retrieve some hand tools, and perform several simple tasks.

Unfortunately, he overexerted himself, his suit heated up, his faceplate fogged over, and as the situation worsened his flailing compounded the problems. He sweated out several pounds of water before finally reentering the capsule.

GEMINI X

LAUNCH DATE July 18, 1966
FLIGHT TIME 2 days, 22 hours, 47 minutes
CREW John Young, Michael Collins

After the failure of the *Gemini IX* EVA, NASA engineers and astronauts modified the EVA suits and developed procedures for undertaking work in weightlessness that would be required to move on to Apollo. John Young and Michael Collins succeeded in docking with the earlier mission's Agena target vehicle. They then maneuvered their docked spacecraft into higher orbit to rendezvous with the drifting Agena left over from *Gemini VIII*.

Collins later commented on the mission patch: "On *Gemini X*, which in my view has the best looking insignia of the Gemini series, artist Barbara Young developed one of John's ideas and came up with a graceful design, an aerodynamic X devoid of names and machines."

GEMINI XI

LAUNCH DATE September 12, 1966
FLIGHT TIME 2 days, 23 hours, 17 minutes
CREW Charles "Pete" Conrad, Richard Gordon

The *Gemini XI* crew docked to the Agena target vehicle only 85 minutes after launch, demonstrating a precision not previously attained by astronauts. Pete Conrad increased the orbit of the docked spacecraft to 850 statute miles, setting a record. Richard Gordon's space walk proved a bit less successful and he had to cut his EVA short. The return to Earth was automatic, a first for the Gemini program, and a precision "splashdown" in the ocean was only 2.8 miles from the rescue ship.

The mission patch shows milestones such as the rendezvous with the Agena, the altitude record, and Gordon's space walk. Conrad and Gordon were both in the Navy, so the patch featured gold thread on a navy blue background.

GEMINI XII

LAUNCH DATE November 11, 1966
FLIGHT TIME 3 days, 22 hours, 35 minutes
CREW James "Jim" Lovell, Edwin "Buzz" Aldrin

This last Gemini flight put in practice the lessons learned on the previous flights. Buzz Aldrin's two-hour tethered space walk demonstrated the feasibility of extravehicular activity. Two later space walks also went smoothly. In all, Aldrin set a record with five hours, 37 minutes of spacewalking. The rendezvous and docking now seemed routine, and the crew performed it "manually," using only the on-board computer and charts when the rendezvous radar failed.

Since the timing for *Gemini XII* was anticipated to be over Halloween in 1966, the crew patch reflected this. Jim Lovell decided on the yellow and black color scheme, and worked with artist Anthony Tharenos at McDonnell Aircraft, the prime contractor for their Gemini spacecraft, to create a patch representing the program's final flight.

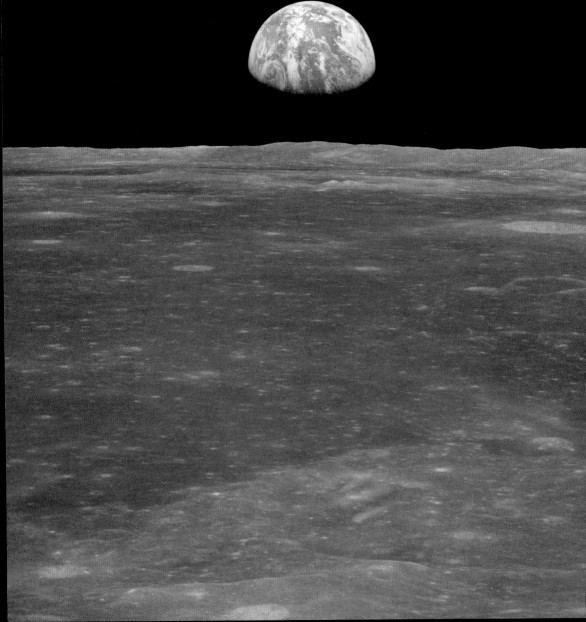

Taken from *Apollo 8* by astronaut Bill Anders, on the first crewed spacecraft to circumnavigate the Moon, *Earthrise* depicted the Earth as a fragile vessel of life. It shows the Moon, gray and lifeless in the foreground, with the blue and white Earth, teeming with life, hanging in the blackness of space. Jim Lovell said, "The vast loneliness is awe-inspiring, and it makes you realize just what you have back there on Earth."

CHAPTER 3
PROJECT APOLLO >>

PROJECT APOLLO

Had there been no Cold War rivalry between the U.S. and the Soviet Union, there would have been no Project Apollo, at least not on the schedule announced by President John F. Kennedy on May 25, 1961. Kennedy's words were immortal: "I believe that this nation should commit itself to achieving the goal, before this decade is out, of landing a man on the Moon and returning him safely to Earth." He was even more eloquent at Rice University during a speech in September 1962: "We choose to go to the Moon! We choose to go to the Moon . . . in this decade and do the other things, not because they are easy, but because they are hard; because that goal will serve to organize and measure the best of our energies and skills, because that challenge is one that we are willing to accept, one we are unwilling to postpone, and one we intend to win, and the others, too."

For the generation of Americans who grew up during the 1960s watching NASA astronauts fly into space, beginning with the Mercury and Gemini programs and culminating with six landings on the Moon, Project Apollo signaled in a very public manner how well the nation could do when it set its mind to accomplishing "hard things," as Kennedy had declared. Television coverage of real space adventures was long and intense, the stakes high, and the risk to life enormous. There were moments of both great danger and high anxiety. In the end, Project Apollo was a triumph of management in meeting enormously difficult systems engineering, and technological and organizational integration requirements.

1]
This was the first test of the S-IB-1, the Saturn IB's first stage. This predecessor of the mighty Saturn V launch vehicle lifted Apollo spacecraft into Earth's orbit, but did not have the thrust required to send spacecraft to the Moon. This test of the first stage at the Marshall Space Flight Center in Huntsville, Alabama, took place on April 13, 1965. Even so, the engine was so powerful that for miles around residents reported the shaking of the ground, the dull roar of the engine, and a few broken windows. It generated a remarkable 1.6 million pounds of thrust when all of its eight H-1 engines fired. This proved nothing when compared to the Saturn V, whose first stage generated 7.5 million pounds of thrust with its five F-1 engines.

2]
This image taken in 1961 shows the first horizontal mating of the components of the Saturn 1B rocket at the Marshall Space Flight Center. In the foreground is the much smaller Redstone rocket, the type of rocket used to launch the first American satellite into orbit on January 31, 1958.

Thereafter, it fell to NASA to organize the plans, marshal the resources, develop the technologies, and carry out these missions. It took a concerted effort throughout, but the first Apollo spacecraft with a crew aboard, *Apollo 7*, flew in the fall of 1968, and the second crew to fly an Apollo craft in space took it to the Moon and back in December 1968. The first Moon landing on July 20, 1969, proved a stunning success. With its completion, NASA could have stopped the program; it had achieved Kennedy's goal, but it persisted. Five more landing missions took place through to the end of 1972, each harvesting a trove of scientific data that has transformed human understanding of the Moon and the solar system.

No technology was more important for the success of Project Apollo than the Saturn V rocket, a massive three-stage vehicle that heaved the Apollo spacecraft, service module, and lunar landing craft to the Moon. Standing 363 feet tall and generating 7.5 million pounds of thrust with its five F-1 rocket engines at liftoff, the Saturn V was a remarkable technological achievement. The Soviet Union tried to build a comparable rocket to reach the Moon during the Space Race, but the N-1 never functioned satisfactorily. All four test launches failed before the Soviets ended their Moon landing program. The Saturn V performed acceptably in every instance.

Equally important, the Apollo space capsule incorporated the knowledge gained during the Mercury and Gemini programs into its design. The vehicle did not fly without flaws. One crew died in a ground test on January 27, 1967, and a second, *Apollo 13*, suffered a nearly deadly accident en route to the Moon in April 1970. Regardless, the spacecraft was an effective vehicle. *Apollo 11* astronaut Michael Collins scrawled inside his spacecraft at the end of the mission: "Spacecraft 107—alias *Apollo 11*—alias *Columbia*. The best ship to come down the line. God Bless Her. Michael Collins, CMP."

The only true spaceship ever built—the lunar module (LM)—could operate only in the vacuum of space or in the one-sixth gravity of the Moon, and nowhere else. An ungainly, spiderlike vehicle, it ferried two members of the crew from lunar orbit to the surface and back flawlessly throughout the program. The *Apollo 10* crew took the spacecraft to within 8.4 nautical miles of the lunar surface while testing the equipment. The crew resisted the desire to go ahead and land. On the *Apollo 13* mission, an explosion in the service module crippled the capsule and the crew used the LM as a lifeboat and made it home safely.

3]
Apollo 11 launched to the Moon on July 16, 1969. Here it passed through the sound barrier—shown by the airflow over the mid-portion of the Saturn V rocket—with a U.S. flag in the foreground.

4]
Apollo 7 crew member Walt Cunningham is bathed in light as he looks out of the porthole during this Earth orbital mission in October 1967.

5]
The *Apollo 9* command module taken from the lunar module Spider on the fifth day of the Earth orbital mission.

3

4

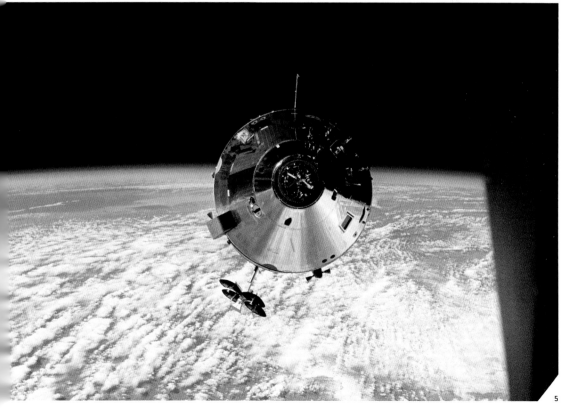

5

By the time of the Apollo missions, NASA had created a formal process for the development and approval of mission patches, as well as for their production and dissemination. In addition, NASA allowed all crews to develop their own mission patches, as well as with *Apollo 9*, to name their spacecraft. The naming of spacecraft had been prohibited during the Gemini program, but it served a practical purpose for Apollo. Since the command and service module (CSM) and LM would need to talk to each other and to Mission Control during flight, the spacecraft names served as a shorthand identifier. *Apollo 9* was the first mission with both vehicles flying; *Apollos 7* and *8* did not have LMs aboard. With NASA leadership's approval, code names for both spacecraft were chosen by the astronauts flying on each mission. The code names were:

- *Apollo 9:* Gumdrop (CSM), Spider (LM)
- *Apollo 10:* Charlie Brown (CSM), Snoopy (LM)
- *Apollo 11:* Columbia (CSM), Eagle (LM)
- *Apollo 12:* Yankee Clipper (CSM), Intrepid (LM)
- *Apollo 13:* Odyssey (CSM), Aquarius (LM)
- *Apollo 14:* Kitty Hawk (CSM), Antares (LM)
- *Apollo 15:* Endeavour (CSM), Falcon (LM)
- *Apollo 16:* Casper (CSM), Orion (LM)
- *Apollo 17:* America (CSM); Challenger (LM)

6]
The last three Apollo landing missions carried a lunar rover to enable the crew to travel more effectively and farther than ever before. *Apollo 17* commander Gene Cernan is at the wheel of the lunar rover at the landing site in December 1972.

7]
Three canopies in the air gently deposit the *Apollo 12* spacecraft in the Pacific Ocean as it returns to Earth with Pete Conrad, Alan Bean, and Dick Gordon following its 1969 lunar landing mission.

8]
Dressed in biological isolation garments, *Apollo 11*'s three-man crew eagerly wait to be picked up by helicopter from the USS *Hornet* after splashing down southwest of Hawaii in the Pacific Ocean. They're joined by a U.S. Navy underwater demolition team swimmer.

6

7

8

In terms of the individual mission patches, it became common practice during the Apollo program for the commander of each mission to fly a T-38 supersonic trainer into the Asheville airport to work with the designers at A-B Emblem—one of the official NASA patch makers—to achieve the vision of the crew. Once the graphic was approved, a drawing would be blown up to exactly six times the size of the patch. The enlargement would be marked with a pencil to show every embroidery stitch required. This would then be fed into a punching machine, which would produce a roll of paper with punches for every stitch. This paper would then be processed through a Swiss embroidery loom to create the mission patches.

The patches used by NASA were sewn onto flight suits, recovery suits, jackets, and any other official NASA gear for the mission. Even ground support personnel had patches affixed to their clothing, sometimes in unusual ways. For example, for each Apollo mission, NASA flight director Gene Kranz's wife made him a new vest to wear in Mission Control. Each vest sported a mission emblem. His cream-colored vest worn for Apollo 13 became famous during the 1995 feature film *Apollo 13*, when actor Ed Harris portrayed Kranz and wore the vest in every scene.

9]
When the crew of *Apollo 11* returned to Earth on July 20, 1969, they were placed in quarantine to ensure that they had not been infected by any lunar germs. NASA used a modified Airstream trailer for this quarantine on the aircraft carrier USS *Hornet*. U.S. president Richard Nixon rushed to the recovery ship to bask in the reflected glory of the *Apollo 11* crew, shown here (from left to right): Neil Armstrong, Michael Collins, and Edwin "Buzz" Aldrin.

10]
Apollo 15 astronaut James Irwin suiting up for this lunar landing mission in the Ready Room at the Kennedy Space Center, July 26, 1971. Clearly seen on his space suit is the mission patch over the left breast.

11]
At the conclusion of the *Apollo 9* Earth orbital mission to test the LM prior to flying it to the Moon, the crew (from left to right) Russell "Rusty" Schweickart, David Scott, and Jim McDivitt walk the red carpet from the helicopter plucking them out of the ocean to a formal ceremony on the deck of the aircraft carrier USS *Guadalcanal*. Their unique mission patches are clearly displayed on the left breasts of their flight suits.

APOLLO 1

LAUNCH DATE None (catastrophic accident, January 27, 1967)

FLIGHT TIME None

CREW Virgil "Gus" Grissom, Roger Chaffee, Edward White

In June 1966, as the crew prepared for a 1967 Apollo spacecraft shakedown mission, NASA approved a mission patch for this first Apollo flight. The crew had worked with Rockwell International artist Allen Stevens to design an appropriate insignia. Stevens would go on to design many other Apollo mission patches for NASA astronauts. Deke Slayton, the head of the Astronaut Office, recalled in his autobiography that since *Apollo 1* "was destined to be the first Apollo orbital flight, this was the prime theme." It featured the Apollo spacecraft orbiting the Earth, with the Moon in the distance. The American flag's star field and red and white stripes formed a circle around the imagery, as well as a gold ring containing the crew's last names and "Apollo 1" in the six o'clock position. The crew began to wear this mission patch on their overalls, flight suits, and space suits during training.

On January 27, 1967, the three-astronaut crew—Gus Grissom, Ed White, and Roger Chaffee—were aboard, running through a mock launch sequence on the launch pad at the Kennedy Space Center. At 6:31 p.m., a fire broke out in the spacecraft. The pressurized high-oxygen environment of the control module quickly fueled flames and toxic smoke. The astronauts died of asphyxiation in a matter of seconds. These were the first fatalities directly linked to America's space program.

This mission and the patch that represents it hold an especially significant place in Project Apollo, even though the mission never took place. Many have noted that the mission's failure, and the loss of the *Apollo 1* crew, revealed many of the serious flaws in the spacecraft. Had it not failed on the ground, it may well have failed in flight to the Moon.

12]
Astronauts for the first Apollo mission (from left to right) Gus Grissom, Ed White, and Roger Chaffee at Launch Complex 34 at Cape Canaveral in early 1967. Notice their mission patches on the right breast of their space suits.

APOLLO 7

LAUNCH DATE October 11, 1968
FLIGHT TIME 10 days, 20 hours, 9 minutes
CREW Walter "Wally" Schirra, Donn Eisele, Walter "Walt" Cunningham

The *Apollo 7* mission proved an enormous confidence builder. It tested in Earth's orbit an extensively redesigned Apollo command module. Wally Schirra, a veteran of the Mercury and Gemini programs, commanded the flight with rookies Donn Eisele and Walt Cunningham as part of the crew. Even though they had their share of problems, the spacecraft performed as intended. But nearly 11 days in orbit took its toll. The crew was overworked, and they all developed colds. As an Earth orbital shakedown, however, *Apollo 7* proved the space-worthiness of the capsule. It also did much more by reenergizing the widespread belief that the Moon landings were possible, and that NASA could accomplish them.

The *Apollo 7* mission patch highlighted the Earth orbital aspects of testing and "human-rating" the Apollo capsule. Cunningham recalled, "It was our original intention to emphasize the *first* manned Apollo (Gus Grissom's flight) and the recovery from the fire on the pad aspects as well. We considered depicting a spacecraft rising from a ball of fire and calling it the Phoenix. The patch designed was subject to NASA approval and we abandoned the Phoenix theme, feeling it would be rejected as in bad taste. I zeroed in on a circle (for the Earth) and an ellipse (for orbit). The orbital plane was tilted for artistic reasons."

13]
The *Apollo* 7 crew checking the command module in the
White Room at the top of the gantry of Launch Complex 39A
at the Kennedy Space Center in May 1968, as they prepared
for their mission. The flight itself did not take place until
October 1968. From left to right: Donn Eisele, Wally Schirra,
and Walt Cunningham. Schirra, a veteran of both the
Mercury and Gemini programs, commanded the mission.

APOLLO 8

LAUNCH DATE December 21, 1968
FLIGHT TIME 6 days, 3 hours, 1 minute
CREW Frank Borman, James "Jim" Lovell, William Anders

It is hard to overstate the significance of the *Apollo 8* circumlunar flight around the Moon in December 1968. NASA had originally envisioned it as a shakedown of key Apollo hardware and procedures in Earth orbit, but national intelligence assessments suggested that the Soviet Union was planning to undertake a human lunar mission in the near term, and transforming *Apollo 8* into a circumlunar flight could steal the march on the Soviets. Moreover, the LM, which was to have been tested by *Apollo 8*, was not ready for flight and performing a repeat of the *Apollo 7* mission seemed unnecessary. NASA replanned the mission and got presidential support.

On December 21, 1968, the mighty Saturn V rocket launched with a crew aboard. Astronauts Frank Borman, Jim Lovell, and William Anders entered lunar orbit on December 24 and remained there until the next day. They gave a memorable Christmas Eve television broadcast, sending holiday greetings to the people of the "Good Earth" before returning home on December 27.

The *Apollo 8* mission patch was elegant in simplicity and unusual in shape. The insignia had the teardrop shape of the command module, and a red figure-eight circling the Earth and Moon represented the number of the mission and the translunar and trans-Earth trajectories.

"The design of the *Apollo 8* patch was quite unique," Jim Lovell recalled. "Borman and I were in California working on our Apollo spacecraft when we got word that our mission had changed. We were going to take McDivitt's spacecraft and make a circumlunar flight around the Moon. On the way back to Houston the next evening, Frank was flying the airplane, and since I had nothing to do, I sort of sketched out what I thought would be an appropriate patch. After I returned to Houston, I gave my sketches to the NASA artist [Gene Rickman], who made the final drawing."

14]
The launch of *Apollo 8* to the Moon on December 21, 1968, the first flight to include astronauts with the Apollo/Saturn V combination.

APOLLO 9

LAUNCH DATE March 3, 1969
FLIGHT TIME 10 days, 1 hour, 1 minute
CREW James "Jim" McDivitt, David Scott, Russell "Rusty" Schweickart

The *Apollo 9* mission served as the first test of the LM under flight conditions, but only in Earth's orbit rather than in the vicinity of the Moon. Launched on March 3, 1969, the crew put the spacecraft through a succession of increasingly sophisticated tests. They repeatedly docked and undocked the two spacecraft, testing the performance of both the LM and the Apollo command module, simulating as closely as possible what future crews would find in lunar orbit. By this time, the astronauts had become proficient at extravehicular activity, but the tandem space walk by Rusty Schweickart and David Scott proved a first. They also tried out the new A7L EVA suit, with Schweickart departing the LM with its self-contained life support backpack while Scott stood up in the hatch.

The *Apollo 9* mission patch, designed by Rockwell International graphic artist Allen Stevens, was based on input from the crew. It emphasized the testing of the lunar module, depicting it flying near the command module. Jim McDivitt noted: "There were 'C,' 'D,' 'E,' 'F,' and 'G' missions . . . each with a certain number of objectives. Our mission was the 'D' mission and the 'D' in McDivitt had a red interior."

APOLLO 10

LAUNCH DATE May 18, 1969
FLIGHT TIME 8 days, 3 minutes
CREW Thomas Stafford, John Young, Eugene "Gene" Cernan

During the *Apollo 10* mission, the crew repeated many of the same hardware tests as had been the case with *Apollo 9*, but for the first time doing so in lunar orbit. A near dress rehearsal for the envisioned *Apollo 11* lunar landing, this mission dramatically flew the LM to within 9.7 statute miles of the surface. The LM performed flawlessly. The altitude reached by the crew was no accident; at that point the descent engine would light to provide maneuverability and landing capability for the next crew. Practicing an approach to the surface during this mission offered valuable lessons for the crew of *Apollo 11*.

Stafford and Cernan flew together on *Gemini IX*, and the patch for this mission was strikingly like that of their earlier flight. Both used a shield as the basic design feature, and focused attention on the spacecraft, the mission objectives, and the large Roman numeral X for the mission number. Once again, Allen Stevens of North American Rockwell (the successor company to Rockwell International) was the artist.

APOLLO 11

LAUNCH DATE July 16, 1969
FLIGHT TIME 8 days, 3 hours, 19 minutes
CREW Neil Armstrong, Michael Collins, Edwin "Buzz" Aldrin

Marking the first human landing on the Moon, this mission made good on the promise of President John F. Kennedy to send a man to the Moon by the end of the decade "and return him safely to the Earth." Six hours after landing on the lunar surface on July 20, astronauts Neil Armstrong and Buzz Aldrin set foot on the lunar surface. Armstrong's words were immortal: "That's one small step for man—one giant leap for mankind." The two spent two and a half hours planting an American flag. But they omitted claiming the land for the U.S. as had been routinely done during European exploration of the Americas. Instead, they unveiled a plaque bearing the inscription: "Here Men from Planet Earth First Set Foot Upon the Moon. July 1969 A.D. We Came in Peace for All Mankind."

The crew and NASA leaders took special care for the design of the *Apollo 11* mission patch. They wanted something elegant that would stand the test of time. Astronaut Jim Lovell suggested featuring a bald eagle on the patch, and Collins found one in a National Geographic book, tracing it over a lunar landscape. Armstrong and Collins also wanted to capture the peaceful nature of the space mission, so they agreed when Tom Wilson, a computer expert and the *Apollo 11* simulator instructor, suggested adding an olive branch to the eagle's beak. NASA artist James Cooper

15]
In this iconic image, Buzz Aldrin stands before the American flag on the lunar surface, July 20, 1969. For a generation of Americans this photograph symbolized the sense of national pride in the success of the Moon landings.

executed the final design that went forward for NASA approval. Dr. Robert Gilruth, director of the Manned Spacecraft Center, refused to authorize the design, however, until the eagle was redrawn to look less ferocious and hostile. The olive branch was placed in the eagle's talons and the expression made a little less menacing. Collins thought the change unnecessary, recalling that the new design made the eagle look "uncomfortable" rather than stately.

APOLLO 12

LAUNCH DATE November 14, 1969
FLIGHT TIME 10 days, 4 hours, 36 minutes
CREW Charles "Pete" Conrad, Richard Gordon, Alan Bean

This second human lunar landing was an exercise in precision. *Apollo 12* landed within 600 feet of the *Surveyor 3* spacecraft in the Ocean of Storms, which had landed two and a half years earlier on April 19, 1967. Pete Conrad's first words on the Moon were exuberant: "Whoopee! Man, that may have been a small one for Neil, but that's a long one for me." Conrad and Bean brought pieces of the *Surveyor 3* back to Earth for analysis, and took two walks lasting just under four hours each.

"The patch was designed by Pete, Dick, and I with the help of about ten other people," recalled astronaut Alan Bean. "The real breakthrough came when a couple of engineers said they thought they could duplicate [our landing site on the Moon] exactly. They came back a few days later and all the craters were the proper size, shape and the lighting was just perfect. They got a relief globe of the moon that was in the library, lit it properly, and then took photographs with a Polaroid at different distances until they got one that had just the right curvature that we wanted on the patch. We selected the blue and gold because they are Navy colors and all of us were in the Navy." Another feature of the patch was that it had four stars, one for each of the crew, and one for C. C. Williams, an astronaut who had been slated for the mission but had died in a T-38 plane crash in late 1967.

APOLLO 13

LAUNCH DATE April 11, 1970

FLIGHT TIME 5 days, 22 hours, 55 minutes

CREW James "Jim" Lovell, Fred Haise, John Swigert

This aborted landing mission turned into one of the most nail-biting rescues in space history. After an explosion 56 hours into the flight, and with power, electrical, and life-support systems failing, NASA engineers quickly turned the lunar module—a self-contained landing craft unaffected by the accident—into a "lifeboat" to provide life support for the return trip. Bringing the crew home alive was now the only objective. It was a close call, but they returned safely on April 17, 1970.

This mission had featured important scientific experiments, which became a feature of the mission patch. Lovell said he "started with the idea of the mythical god, Apollo, driving his chariot across the sky and dragging the sun with it. We eventually gave this idea to an artist named Lumen Winter, and he came up with the three-horse design that symbolized the Apollo."

APOLLO 14

LAUNCH DATE January 31, 1971
FLIGHT TIME 9 hours, 2 minutes
CREW Alan Shepard, Stuart Roosa, Edgar Mitchell

Apollo 14 undertook a mission similar to what the aborted *Apollo 13* flight had attempted and landed in the highlands of Fra Mauro. Alan Shepard and Edgar Mitchell undertook two walks on the lunar surface, deployed broader experiments to the landing site, and using what could be called a "lunar rickshaw" to transport their equipment, they traveled farther than any earlier crew to collect samples. This mission represented a major step forward in terms of harvesting knowledge about the Moon.

The crew of *Apollo 14* employed an oval design for their patch, which is chiefly remarkable for the odd choice of lettering used. It uses the astronaut logo as the central feature; astronauts received a silver lapel pin with this emblem upon acceptance into the astronaut corps and a gold version after their first spaceflight. NASA artist Jean Beaulieu executed the design based on the astronauts' ideas.

APOLLO 15

LAUNCH DATE July 26, 1971
FLIGHT TIME 12 days, 17 hours, 12 minutes
CREW David Scott, Alfred Worden, James "Jim" Irwin

Apollo 15 represented the first of a series of missions that carried a lunar rover. This made the Moon's most interesting surface features, its mountains and rilles, more accessible. The most important aspect of this was the expansion of the Apollo lunar surface experiments package, a set of instruments used by the astronauts to measure soil mechanics, meteoroids, lunar ranging, magnetic fields, and solar wind experiments. Exploring the Hadley Rille, David Scott and James Irwin discovered a sample of ancient lunar crust nicknamed the "Genesis Rock," which helped determine the origins of the Moon.

According to Apollo 15 astronaut Alfred Worden, "The mission patch for Apollo 15 was basically designed by the Italian dress designer Emilio Pucci. We took his design, changed it from a square to a circular patch, made it red, white, and blue, and put a lunar background behind the three stylized birds that were the major Pucci contribution. The symbology is of three stylized birds flying over the lunar surface." With the design solidified, NASA artist Jerry Elmore executed the final version of the mission patch.

APOLLO 16

LAUNCH DATE April 16, 1972

FLIGHT TIME 11 days, 1 hour, 51 minutes

CREW John Young, Charles Duke, Thomas Mattingly

This fifth human lunar landing took place in the Descartes highland region, with the astronauts spending more than three days on the lunar surface. The crew collected numerous rock and soil specimens, including a 24-pound chunk that gained fame as the largest single rock returned by the Apollo astronauts. John Young and Charlie Duke also revved up their lunar rover, at one time getting up to a top speed of 10.94 miles per hour.

Duke told how the *Apollo 16* mission patch came to be in his autobiography, *Moonwalker*: "John, Ken, and I had several basic ideas we wished to incorporate to commemorate our mission: patriotism, teamwork, and the Moon. To show teamwork, the yellow NASA wishbone symbol of flight was placed on top of the seal... circling a blue and gold border were our names... and 16 white stars to emphasize outer space and the number of our flight."

APOLLO 17

LAUNCH DATE December 7, 1972
FLIGHT TIME 12 days, 13 hours, 52 minutes
CREW Eugene "Gene" Cernan, Harrison Schmitt, Ronald Evans

The sixth and final Apollo human lunar landing significantly extended the time spent on the Moon. It once again made use of a roving vehicle and its crew included the only geologist, Harrison Schmitt, to set foot on the Moon. They collected a record 239.49 pounds of rocks. They also used the lunar rover to travel 20.96 statute miles, discovering a unique orange-colored soil that helped determine the origins of the Moon. With this mission, the lunar landing program ended.

Famed space artist Robert T. McCall designed the mission patch for the *Apollo 17* crew. Mission commander Gene Cernan stated in his autobiography, *The Last Man on the Moon:* "A deep blue background featured the Moon, Saturn, and a spiral galaxy, with the eagle's wing just touching the Moon to suggest that this celestial body had been visited by man. Apollo gazes to the right toward the galaxy to imply further exploration, with the eagle leading mankind into the future."

This image of *Skylab* in orbit was taken as the third crew (*Skylab*-4) departed the space station after 84 days in the orbiting laboratory.

CHAPTER 4
PROJECT SKYLAB AND ASTP »

PROJECT SKYLAB AND ASTP

At the end of the Apollo program, NASA considered alternative uses for the Apollo/Saturn hardware that had already been produced. Ideas abounded, but the most readily possible was to build an orbital workshop out of a third stage of a Saturn V rocket and crew it with astronauts flown there in the Apollo capsule. This became known as *Skylab*, a prototype space station. Launched on May 14, 1973, in a mission that also marked the last flight of the giant Saturn V rocket, *Skylab* almost immediately faced severe malfunctions. Vibrations during liftoff shook the meteoroid shield—designed to protect the workshop from damage in orbit as well as to shade *Skylab* from the Sun's rays—off the rocket, taking with it one of the workshop's two solar panels.

Despite this accident, the station successfully achieved orbit at the desired altitude of 270 miles. NASA's mission control maneuvered *Skylab* so the remaining solar panels faced the Sun, compensating as much as possible for the loss of the main solar panel. The first crew to reach *Skylab*—the crew called it *Skylab 1*—launched on May 25, 1973, with plans to make the workshop habitable. After doing so they stayed aboard until June 22, 1973. Two additional crews occupied *Skylab* until February 8, 1974. Altogether the three crews on *Skylab* worked for a total of 171 days and 13 hours, during which time it was the site of nearly 300 scientific and technical experiments.

1]
This artist's concept of the *Skylab* space station shows astronauts in the orbital workshop.

2]
An important feature of *Skylab* was that it offered room in the forward dome to test equipment. *Skylab 3* commander Alan Bean tested the prototype M509 astronaut maneuvering unit throughout his time aboard. He usually wore his coveralls for these tests, but in this image from 1973 he is in a full EVA suit. The backpack featured maneuvering jets using compressed air and hand controllers easily accessed by the astronaut. This backpack was the predecessor of the manned maneuvering unit developed and used during the space shuttle era.

3]
The long-duration *Skylab* missions needed to ensure that the astronauts could live and work in a weightless environment over several months. NASA developed several technologies to ensure the health of the crews. One was a lower-body negative pressure device to simulate the stress on the cardiovascular system experienced in normal gravity. In preparation for the *Skylab 2* mission during a simulation at NASA's Johnson Space Center, astronaut Paul Weitz tests the device while astronaut Joseph Kerwin oversees the tests.

1

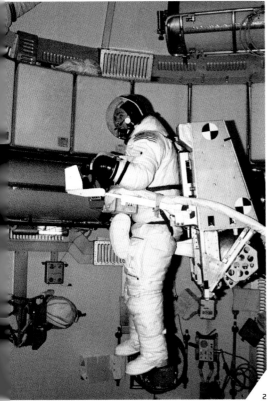

2

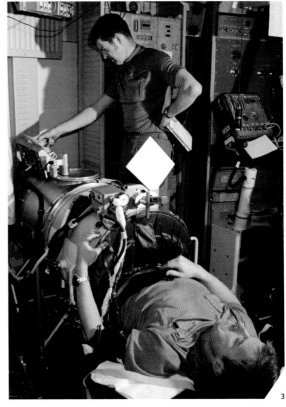

3

With the departure of the last *Skylab* crew, NASA shut down the station's systems. It went into a rapidly decaying orbit in the late 1970s and returned to Earth on July 11, 1979, scattering a trail of debris across a chunk of sparsely populated land in Western Australia.

The last flight of any Apollo hardware took place in July 1975 during a period of improved relations between the U.S. and the Soviet Union. It was the first international human spaceflight mission and was specifically planned to test the compatibility of rendezvous and docking systems for American and Soviet spacecraft in the hope that developing such a capability might open the way for future joint missions. NASA engineers designed and built a universal docking module that would serve as an airlock and transfer corridor between the two nations' spacecraft.

While NASA's missions had been more focused on pure scientific research since the end of the Moon landings, the *Apollo-Soyuz* flight was primarily diplomatic. The two missions both launched on July 16, 1975, and the two craft docked in orbit two days later as the two crews undertook joint experiments and exchanged diplomatic greetings and gifts. Although there were no similar joint missions for the rest of the Cold War, the test project did serve to lead the way for greater international cooperation in human spaceflight in the years that followed.

As with the previous human spaceflights, the astronauts designed patches for their missions. The official numbering of *Skylab* missions has the launch of the orbital workshop as *Skylab 1*, and the three subsequent crew launches were designated *Skylab 2, 3,* and *4*. An unofficial numbering scheme denoted the three crew launches as *Skylab 1, 2,* and *3*, leaving the launch of the orbital workshop undesignated. The crew patches followed the unofficial numbering scheme.

4]
This artist's concept shows the first international docking of the U.S. Apollo spacecraft with the USSR's Soyuz spacecraft in July 1975.

5]
Skylab in orbit at the end of its mission in 1974.

SKYLAB 1

LAUNCH DATE May 25, 1973
FLIGHT TIME 28 days, 50 minutes
CREW Charles "Pete" Conrad, Joseph Kerwin, Paul Weitz

The *Skylab 1* crew lifted off from Kennedy Space Center and rendezvoused with the workshop. After substantial extravehicular repair work, including the deployment of a large space parasol sunshade that cooled the inside to a more manageable 75°F, *Skylab* became both habitable and operational. The crew went on to conduct solar astronomy, medical studies, and various scientific experiments on the space station over the course of 404 orbits before returning to Earth on June 22, 1973, leaving the space station unoccupied until the arrival of the *Skylab 2* crew on July 28, 1973.

The *Skylab 1* mission patch was designed by famed artist Frank Kelly Freas. Known as the "Dean of Science Fiction Artists," Freas emphasized *Skylab* in the design, silhouetted against the Earth's globe. This, in turn, is shown eclipsing the Sun, with the brilliant signet-ring pattern of the moment before total eclipse.

SKYLAB 2

LAUNCH DATE July 28, 1973
FLIGHT TIME 59 days, 11 hours, 9 minutes
CREW Alan Bean, Jack Lousma, Owen Garriott

The second astronaut mission to *Skylab* stretched the mission as never before. The crew—Alan Bean, Jack Lousma, and Owen Garriott—spent 59 days in space, an American record at the time. The activities were generally mundane, allowing the crew to concentrate on research in microgravity of many different types. They experimented with prototype equipment envisioned for use in future missions to the Moon and planets. They tended the central scientific instrument onboard *Skylab*, the solar telescope, which used film that had to be changed during space walks. This instrument yielded a wealth of new scientific data about the Sun and its corona. In life sciences, the crew focused on recording how their bodies responded to microgravity and tested how a spider performed. They found that, regardless of weightlessness, the spider could still spin webs.

Artists at McDonnell Douglas adapted a Leonardo da Vinci sketch for this misison patch. Jack Lousma recalled, "We had decided that our patch should be red, white, and blue for obvious reasons . . . the Sun half is a little special in that the solar flare depicted in yellow-orange is the shape of one Owen Garriott had done extensive analysis on years before. Leonardo da Vinci's man represents the medical aspects of the flight."

SKYLAB 3

LAUNCH DATE November 16, 1973
FLIGHT TIME 84 days, 1 hour, 16 minutes
CREW Gerald Carr, Edward Gibson, William Pogue

This crew docked with *Skylab* in the longest-duration mission, and the last of the Skylab program. Carrying out a grueling work pace that caused some tension with Mission Control, the crew observed Comet Kohoutek, a solar eclipse, and solar flares. The astronauts conducted four space walks, including one on Christmas Day to view Kohoutek.

With the intensive scientific agenda of this flight, the crew wanted to reflect that in its mission patch. As designed by Barbara Matelski from the Johnson Space Center's graphics department, the *Skylab 3* mission patch symbolized, according to the crew, "the three major areas of investigation in the mission." It had three major elements: a tree that represented the natural environment, a hydrogen atom that is the most common element known in the universe, and the human body emphasizing the life sciences aspects of this flight. Additionally, the crew remarked that the hydrogen atom also recognizes the solar investigations of *Skylab*, as the Sun is largely composed of hydrogen.

APOLLO-SOYUZ TEST PROJECT (ASTP)

LAUNCH DATE July 15, 1975
FLIGHT TIME 9 days, 7 hours, 28 minutes
CREW Thomas Stafford, Donald "Deke" Slayton, Vance Brand

The ASTP was a diplomatic effort to assuage tensions between the U.S. and the Soviet Union in the Cold War. Directed by the Nixon administration as it undertook a détente in the U.S./USSR rivalry, it called for a joint rendezvous and docking mission in Earth's orbit. The close collaboration between NASA and its Soviet counterpart involved the building of a universal docking mechanism so that the Apollo and Soyuz spacecraft could link up in orbit. The flight was a stunning success as a goodwill mission, although little in terms of scientific value resulted from it, nor did it foster additional joint missions.

The ASTP patch was designed by Jean Pinataro of North American Rockwell, the prime contractor for the Apollo spacecraft. Initially unapproved as being insufficiently international in focus, the patch was redesigned by Pinataro with a central vignette derived from Robert McCall's 1974 painting of the orbital rendezvous. She later said that the crew involved themselves in every aspect of this redesign. "I recall being annoyed that the astronauts' directions were so explicit that I was unable to connect the three elements in that central area." But it worked, and in December 1974 the redesigned patch was approved.

This image of space shuttle *Challenger* in space, taken by the Shuttle Pallet Satellite (SPAS) shows the cargo bay open and the remote manipulator system extended.

CHAPTER 5
SPACE SHUTTLE
1981–1986

SPACE SHUTTLE 1981–1986

When NASA began work on what became known as the space shuttle in the late 1960s, few recognized how important a part of American life the program would become through its 30-year flight history. The space shuttle consisted of three primary elements: a delta-winged orbiter, two solid rocket boosters, and an external fuel tank. The orbiter and the two solid rocket boosters were reusable. The shuttle was designed to transport approximately 45,000 tons of cargo into near-Earth orbit, 115 to 250 statute miles above the Earth. It could also accommodate a large flight crew for more than a week in space.

After a decade of research and development, with much public excitement, *Columbia*, the first orbiter that could reach space, took off from the Kennedy Space Center on April 12, 1981, successfully landing nearly two days later. After the first four flights, President Ronald Reagan declared the system operational. Thereafter it would be used for all government launches, and NASA sought to sell rides into orbit for commercial satellites.

1]
The first launch of the space shuttle *Columbia* took place on April 12, 1981, 20 years to the day after the first human, Soviet cosmonaut Yuri Gagarin, reached space. The reusable space shuttle orbiter is strapped to a white painted external tank with a solid rocket booster on each side. It has just cleared the launch tower at the Kennedy Space Center, the point at which the launch controllers hand over responsibility to the Mission Control Center at the Johnson Space Center in Houston.

2]
Astronaut Sally Ride, pictured here during the flight of STS-7 in 1983, was the first American woman to fly into space.

3]
Astronaut Bruce McCandless undertook a series of important space walks during the STS-41B mission in 1984. He is shown here attached by his feet to the Canadarm remote manipulator arm while working in the shuttle's payload bay. During this mission, McCandless also performed the first test of the manned maneuvering unit, which allowed untethered movement during an EVA.

There were four reusable space shuttle orbiters in the fleet for most of the program's life. The first, *Columbia*, was lost in a tragic accident in 2003. The second orbiter, *Challenger*, debuted in 1983 but was lost in a horrific launch accident in 1986. *Discovery* became the third orbiter in 1984, and *Atlantis* began flying in 1985. A replacement orbiter, *Endeavour*, joined the fleet in 1991.

Despite high hopes, although the system was reusable, its complexity meant that the turnaround time between flights was several months instead of several days. In addition, missions were delayed for a wide range of problems associated with the safety and performance of the vehicle.

Like earlier human spaceflight programs, the shuttle crews designed their own mission patches, reflecting the major elements of the flight and the names of the crew. Each became an individualized statement of crew activity.

4]
The crew of STS-41G during the walk to the van for transport to the space shuttle *Challenger* on the morning of October 4, 1984. Sally Ride was making her second trip into space, while Kathryn Sullivan was making her first. Sullivan became the first woman to undertake an EVA. The picture shows on the left (from front to back) Kathryn Sullivan, mission commander Robert Crippen, Paul Scully-Power, and Jon McBride, and on the right (from front to back) Sally Ride, David Leestma, and Marc Garneau.

5]
The launch of STS-41G in the early hours of October 5, 1984, offered a stunning vision for all who saw it. The space shuttle *Challenger* climbed toward orbit as its rocket engine lit up the dawn sky.

6]
During the early years of the space shuttle program, several missions landed at Edwards Air Force Base in Southern California. Situated in a Mojave Desert salt pan, the base's miles of landing strip made it an easy target for the early flights. Here the crew of STS-61A touches down on November 6, 1985.

STS-1 (COLUMBIA)

LAUNCH DATE April 12, 1981
FLIGHT TIME 2 days, 6 hours, 21 minutes
CREW John Young, Robert Crippen

This was the first flight of space shuttle *Columbia*, and the first landing of an airplane-like craft from orbit that could be reused. In space, the crew tested *Columbia*'s on-board systems; fired the orbital maneuvering system used for changing orbits and the reaction control system engines used for altitude control; opened and closed the payload bay doors (the bay was empty for this first flight); and, after 36 orbits, made a smooth touchdown at Edwards Air Force Base.

The patch for STS-1 was designed by famed space artist Robert McCall. It represents both the first shuttle launch and orbit of the Earth. There are subtle differences between the patch made for the STS-1 crew and the versions created for sale in the memorabilia market. For example, the flames below the space shuttle are embroidered in four distinct colors of thread on the crew-worn patch, whereas a version available for public sale has a simple two-color flame.

STS-2 (COLUMBIA)

LAUNCH DATE November 12, 1981
FLIGHT TIME 2 days, 6 hours, 13 minutes
CREW Joe Engle, Richard Truly

During the second flight of space shuttle *Columbia*, the crew continued to test the systems aboard. They also checked out the Canadarm remote manipulator arm, essentially a crane that could be deployed from the payload bay and controlled by astronauts inside the spacecraft to capture, turn, hoist, and move bulky satellites and other objects. The Canadarm would become an indispensable tool used in many missions. The crew also tended the first scientific payload on the shuttle, a set of experiments carried in the shuttle's mid-deck. The STS-2 mission patch is reminiscent of the *Apollo 17* mission patch with the eagle and flag design, this time with the shuttle sailing into orbit.

STS-3 (COLUMBIA)

LAUNCH DATE March 22, 1982
FLIGHT TIME 8 days, 5 minutes
CREW Jack Lousma, Gordon "Gordo" Fullerton

The longest of the shuttle orbital test flights, STS-3 carried a second scientific payload. It also included the second test of the remote manipulator arm. The flight duration was extended by one day because of flooding at the primary landing site, and an alternative landing site had to be used. The mission patch for STS-3 employed a unique sunburst design, with the shuttle in orbit facing forward. The remote manipulator arm is featured on the patch.

STS-4 (COLUMBIA)

LAUNCH DATE June 27, 1982
FLIGHT TIME 7 days, 1 hour, 10 minutes
CREW Thomas "Ken" Mattingly, Henry "Hank" Hartsfield

This was the fourth flight of STS, which included the first Department of Defense payload, plus additional scientific payloads. The testing program was completed during the flight and the shuttle was returned for reuse. The oval shape of the mission patch emphasizes the shuttle streaking across its orbit. Again, there were subtle differences between the crew's patch and those available for sale. The version worn by the astronauts had a small area of lighter red thread in the flag stripes following the space shuttle.

STS-5 (COLUMBIA)

LAUNCH DATE November 11, 1982
FLIGHT TIME 5 days, 2 hours, 14 minutes
CREW Vance Brand, Robert Overmyer, Joseph Allen, William Lenoir

President Ronald Reagan declared the space shuttle operational after the fourth flight, and this was reflected in STS-5. It carried two commercial satellites (*SBS 3* and *Anik C3*) that were deployed into orbit from the shuttle's payload bay. These were removed by the Canadarm and then deployed to their appropriate orbits using the payload assist module upper stage. The four crew members aboard were to undertake an EVA test of new space suits, but this had to be canceled when two suits malfunctioned. The patch for this fifth space shuttle mission emphasizes the deployment of the two commercial satellites as the module launches them to a different orbit.

STS-6 (CHALLENGER)

LAUNCH DATE April 4, 1983

FLIGHT TIME 5 days, 24 minutes

CREW Paul Weitz, Karol Bobko, Story Musgrave, Donald Peterson

STS-6 was space shuttle *Challenger*'s debut flight, and it included the shuttle program's first space walks. Since the STS-5 EVA had to be canceled, astronauts Story Musgrave and Donald Peterson tested the new space suits during this mission. Their four-hour space walk was the first for any American astronaut since the *Skylab* missions. This mission also deployed *TDRS-1*, the first of NASA's tracking and data relay satellites, which is depicted in the mission patch design. Additionally, this sixth space shuttle flight is represented by the hexagonal shape of the mission patch, and the six stars in the constellation Virgo.

STS-7 (CHALLENGER)

LAUNCH DATE June 18, 1983

FLIGHT TIME 6 days, 2 hours, 24 minutes

CREW Robert Crippen, Frederick "Rick" Hauck, John Fabian, Sally Ride, Norman Thagard

In this, the seventh flight of STS, the crew rendezvoused with and retrieved the German-built shuttle pallet satellite experiment platform, which took the first full pictures of a shuttle orbiter in space. They also released two communication satellites into orbit. The five crew members made it the largest crew in the shuttle program to date and included the first female U.S. astronaut, Sally Ride. The mission patch shows the orbiter circling Earth, with the remote manipulator arm forming the number 7, with seven stars visible against the black sky, and five arms within the Sun representing the number of crew members.

STS-8 (CHALLENGER)

LAUNCH DATE August 30, 1983
FLIGHT TIME 6 days, 1 hour, 9 minutes
CREW Richard Truly, Daniel Brandenstein, Dale Gardner, Guion "Guy" Bluford, William Thornton

This mission launched the commercial satellite *Insat IB* from space shuttle *Challenger*. This was Guy Bluford's first flight and also marked the first time an African American person had traveled to space. STS-8 featured the program's first night launch and landing. Physician William Thornton conducted biomedical experiments. The mission patch depicts the shuttle from a different angle, and shows the solid rocket boosters separating after launch. It also shows the *Challenger*'s launch at night. The eighth flight of the space shuttle program is represented by the eight stars of the constellation Aquila.

STS-9 (COLUMBIA)

LAUNCH DATE November 28, 1983
FLIGHT TIME 10 days, 7 hours, 47 minutes
CREW John Young, Brewster Shaw, Owen Garriott, Robert Parker, Ulf Merbold, Byron Lichtenberg

STS-9 involved the first deployment of European-built *Spacelab 1*, a multidisciplinary experiments module intended to demonstrate that good science could be done on short shuttle flights. There were 73 experiments aboard. STS-9 was actually a mission of numerous "firsts"—it was the first mission with six crew members (the biggest crew to date), and the first to include non-U.S. astronauts, one of whom was West German, to fly in a U.S. space program. The patch features the *Spacelab 1* module flown in the payload bay, while the nine stars and the path of the orbiter represent the flight's number in the space shuttle program.

STS-41B (CHALLENGER)

LAUNCH DATE February 3, 1984

FLIGHT TIME 7 days, 23 hours, 16 minutes

CREW Vance Brand, Robert "Hoot" Gibson, Ronald McNair, Robert Stewart, Bruce McCandless

This mission continued NASA's efforts to provide commercial satellite deployment services for paying customers. *Palapa B-2* and *Westar V-I* were both commercial communications satellites deployed during this mission. Unfortunately, neither reached its appropriate orbit and so both had to be retrieved and redeployed. Additionally, *Challenger* made its first landing at NASA's Kennedy Space Center, after the building of an extended runway that could accommodate the shuttle. For this flight, NASA unveiled its new space shuttle numbering system instead of using a sequential delineation. After "STS-" the number stood for the last number of the year the mission was to be flown. In this mission, therefore, the 4 stood for 1984. NASA and the U.S. Air Force proposed using two launch sites for the shuttle with the Kennedy Space Center represented by 1 and a launch site at Vandenburg Air Force Base in California designated 2. The alphabetical designator was the flight number of that year. This numbering system was abandoned after the *Challenger* accident in 1986. The mission patch was designed by Robert McCall. On the left is an illustration of a satellite deployment, and on the right an astronaut makes an untethered EVA.

STS-41C (CHALLENGER)

LAUNCH DATE April 6, 1984

FLIGHT TIME 6 days, 23 hours, 40 minutes

CREW Robert Crippen, Francis "Dick" Scobee, Terry Hart, James van Hoften, George Nelson

This shuttle flight deployed a long-duration exposure facility (LDEF), which tested the effects of space exposure on different materials. George Nelson and James van Hoften replaced a faulty altitude control system and one science instrument on the *Solar Maximum* satellite, and then re-released it into orbit. The mission patch for STS-41C was especially complex, reflecting the major elements of the mission in the visor of a space suit helmet. It shows *Challenger* and its remote manipulator system deploying the LDEF, the Earth and blue sky, and an astronaut working at the damaged *Solar Maximum* satellite.

STS-41D (DISCOVERY)

LAUNCH DATE August 30, 1984

FLIGHT TIME 6 days, 56 minutes

CREW Henry "Hank" Hartsfield, Michael Coats, Richard Mullane, Steven Hawley, Judith Resnick, Charles Walker

This was the first flight of *Discovery*. NASA continued its commercial satellite deployment with the release of three communications satellites. This mission was also a first for a new NASA program payload specialist. This position was for commercially employed astronauts to fly and tend to commercial activity on the space shuttle. Many payload specialists would fly before NASA ended the program in 1986. This mission patch features the *Discovery*, NASA's third orbital vehicle, as it makes its maiden voyage. The sailing ship on the left represents the orbiter's namesake, which figured prominently in the history of exploration.

STS-41G (CHALLENGER)

LAUNCH DATE October 5, 1984

FLIGHT TIME 8 days, 5 hours, 24 minutes

CREW Robert Crippen, Jon McBride, Kathryn Sullivan, Sally Ride, David Leestma, Paul Scully-Power, Marc Garneau

STS-41G was the first seven-member crew and included Kathryn Sullivan, the first woman to walk in space. The mission patch focuses on the American flag and a symbol known as the astronaut emblem. It shows a trio of trajectories merging into a bright star encircled by an elliptical wreath. This was the first instance of payload specialists being added to the outer part of a mission patch, a method of delineation that continued until STS-35 in 1990. In this case, Garneau and Scully-Power were added to the mission after the patch had been designed.

STS-51A (DISCOVERY)

LAUNCH DATE November 8, 1984

FLIGHT TIME 7 days, 23 hours, 45 minutes

CREW Frederick "Rick" Hauck, David Walker, Joseph Allen, Dale Gardner, Anna Fisher

This mission delivered two satellites into orbit, before bringing back two others whose onboard boosters had failed. The mission patch, designed by Maryland artist Stephen Hustvedt, depicts *Discovery* as a soaring eagle, en route to orbit Earth. The red and white trailing stripes and blue background along with the presence of the eagle resonate with patriotism. The two satellites orbiting the Earth in the background amidst a celestial scene are a universal representation of the versatility of the space shuttle.

STS-51C (DISCOVERY)

LAUNCH DATE January 24, 1985
FLIGHT TIME 3 days, 1 hour, 33 minutes
CREW Thomas "Ken" Mattingly, Loren Shriver, Ellison Onizuka, James Buchli, Gary Payton

This mission supported the Department of Defense and the mission remains classified. The patch reflects the mission's classified status in a design that's reminiscent of a U.S. Air Force medal. This mission patch only exists with a vacuum-sealed plastic backing. In some versions, the gold border is extended around the eagle's wing tips while on others, the space shuttle in the center has a distinctive band of thick gray thread running along its top edge.

STS-51D (DISCOVERY)

LAUNCH DATE April 12, 1985
FLIGHT TIME 6 days, 23 hours, 55 minutes
CREW Karol Bobko, Donald Williams, Rhea Seddon, David Griggs, Jeffrey Hoffman, Charles Walker, Jacob Garn

In STS-51D, *Discovery*'s first official flight, the crew retrieved a communications satellite, and Utah senator Jacob Garn became the first member of Congress to fly in space. The dominant feature of the mission patch is an orbit formed by *Discovery* and the colonial American flag. The flag in orbit signifies the U.S. presence in space and preeminence in human spaceflight. The original 13-star flag is used to symbolize a continuity of technical achievement and progress since colonial times. Positioning the name *Discovery* in front of the flag represents the U.S. spirit of discovery and exploration of new frontiers.

STS–51B (CHALLENGER)

LAUNCH DATE April 29, 1985
FLIGHT TIME 7 days, 9 minutes
CREW Robert Overmyer, Frederick Gregory, Don Lind, Norman Thagard, William Thornton, Taylor Wang, Lodewijk van den Berg

In this mission, *Spacelab 3*, in *Challenger's* large cargo bay, housed 15 experiments on topics including materials processing, fluid behavior, atmospheric physics, astronomy, and life sciences. The patch for this mission was designed by NASA artist Carol Ann Lind. It depicts *Discovery* and its science module payload. The seven stars of the constellation Pegasus surround the orbiting spacecraft above a flag-draped Earth.

STS–51G (DISCOVERY)

LAUNCH DATE June 17, 1985
FLIGHT TIME 7 days, 1 hour, 39 minutes
CREW Daniel Brandenstein, John Creighton, John Fabian, Steven Nagel, Shannon Lucid, Sultan Abdul Aziz Al-Saud, Patrick Baudry

Three communications satellites were deployed during this flight, as well as one reusable payload. It was the first U.S. flight with French and Saudi Arabian crew members. Depicting the space shuttle as the *Wright Flyer* of the Space Age, the design calls attention to the advances in aviation technology in the U.S. within a relatively short span of the 20th century. The international nature of the flight is emphasized, with the flags of the French and Saudi Arabian payload specialists appearing next to their names at the bottom.

STS-51F (CHALLENGER)

LAUNCH DATE July 29, 1985

FLIGHT TIME 7 days, 22 hours, 45 minutes

CREW Gordon "Gordo" Fullerton, Roy Bridges, Karl Henize, Story Musgrave, Anthony England, Loren Acton, John-David Bartoe

This *Spacelab 2* mission focused on astronomical investigations, especially concerning the Sun. The shuttle payload bay contained three pallets of scientific instruments, specifically designed to measure plasma physics, high-energy astrophysical phenomena, and the physics of the Sun and cosmic rays. NASA artist Skip Bradley at the Johnson Space Center depicted the space shuttle *Challenger* at a tilted angle rising toward a bright star in the distance. Since the *Spacelab 2* instruments focused on astronomy, the background features a range of stars, including the constellations Leo and Orion.

STS-51I (DISCOVERY)

LAUNCH DATE August 27, 1985

FLIGHT TIME 7 days, 2 hours, 18 minutes

CREW Joe Engle, Richard Covey, James van Hoften, John "Mike" Lounge, William Fisher

The crew on this mission rescued the *Syncom IV-3* satellite (also known as "Leasat"), stranded in orbit on STS-51D, and repaired and reboosted it through two space walks by James van Hoften and William Fisher. The mission patch makes a patriotic statement with the basic colors of red, white, and blue, suggesting the American flag, and a dominant American bald eagle in flight. The shock wave represents those formed by the orbiter during the reentry phase of the flight to Earth.

STS-51J (ATLANTIS)

LAUNCH DATE October 3, 1985

FLIGHT TIME 4 days, 1 hour, 45 minutes

CREW Karol Bobko, Ronald Grabe, David Hilmers, Robert Stewart, William Pailes

The first flight of space shuttle *Atlantis* was a classified mission for the Department of Defense. Officially, no information about the activities on the flight were released, but some speculated that one aspect of the mission involved the deployment of at least one surveillance satellite. This mission patch offers a patriotic take on the activities of its crew in the larger scheme of national defense. There is an emphasis on the Statue of Liberty as a symbol of American freedom. For more about the classified missions, see page 114.

STS-61A (CHALLENGER)

LAUNCH DATE October 30, 1985

FLIGHT TIME 7 days, 45 minutes

CREW Henry "Hank" Hartsfield, Steven Nagel, Bonnie Dunbar, James Buchli, Guion Bluford, Reinhard Furrer, Ernst Messerschmid, Wubbo Ockels

The Federal Republic of Germany sponsored the *Spacelab D-1* payload flying on this flight, tended by German astronaut Ernst Messerschmid. It had experiments aboard relating to fluid physics, material science, biology, medicine, and navigation. A special emphasis in the experiments package involved weightlessness and human physiology. The mission patch represents the surnames of the crew of eight (a record) surrounding the *Challenger* carrying a science module and colorful flags representing an international crew from the U.S. and Germany.

STS-61B (ATLANTIS)

LAUNCH DATE November 26, 1985
FLIGHT TIME 6 days, 21 hours, 5 minutes
CREW Brewster Shaw, Bryan O'Connor, Jerry Ross, Mary Cleave, Sherwood "Woody" Spring, Charles Walker, Rodolfo Neri Vela

This mission deployed three communications satellites and carried out the EASE/ACCESS experiment, which centered around assembling large structures during space walks. Appropriately patriotic, the mission patch uses flag and rainbow symbolism to highlight the shuttle in orbit. The design is surrounded by the surnames of the seven crew members. Astronaut Rodolfo Neri Vela has the flag of Mexico next to his name in honor of his country of birth.

STS-61C (COLUMBIA)

LAUNCH DATE January 12, 1986
FLIGHT TIME 6 days, 2 hours, 4 minutes
CREW Robert "Hoot" Gibson, Charles Bolden, George Nelson, Steven Hawley, Franklin Chang-Diaz, Robert Cenker, William "Bill" Nelson

This mission is largely memorable because of Florida congressman Bill Nelson flying on it. In addition, the crew deployed a communications satellite for the RCA company and conducted several smaller experiments. On the mission patch, *Columbia*, which opened the shuttle era with four orbital flight tests in 1981–82, is shown during reentry. Gold lettering of the astronauts' names swirls through the patch, with a shock-wave pattern encircling the orbiter. The payload specialists are named outside the patch proper.

STS-51L (CHALLENGER)

LAUNCH DATE January 28, 1986

FLIGHT TIME 1 minute, 13 seconds

CREW Francis "Dick" Scobee, Michael Smith, Ellison Onizuka, Judith Resnick, Ronald McNair, Sharon Christa McAuliffe, Gregory Jarvis

In this tragic mission, all seven members of the crew were lost 73 seconds into the flight when the *Challenger* disintegrated after the solid rocket boosters ignited the external tank. Several investigations followed the accident, the most important being the presidentially mandated commission chaired by William Rogers. It found that the *Challenger* accident resulted from a poor engineering decision, an O-ring used to seal joints in the solid rocket booster that was susceptible to failure at low temperatures. The Rogers Commission also criticized the communication system within NASA, finding that the potential for O-ring failure had been understood by NASA engineers prior to the launch of *Challenger* and that the accident could have been avoided. This tragic mission ended the first phase of the space shuttle program.

The STS-51L mission patch depicts the shuttle launching from Florida and soaring into space. It includes some of the prescribed duties of the crew—observation and photography of Halley's Comet—backdropped against the American flag. Among the crew names, the name of the first teacher in space, Sharon Christa McAuliffe, is followed by a symbolic apple for a teacher.

The space shuttle *Atlantis* launches on December 2, 1988. STS-27 was the second mission after the *Challenger* accident. Its astronauts undertook a dedicated Department of Defense mission.

CHAPTER 6
SPACE SHUTTLE
1988–1993
« »

SPACE SHUTTLE 1988–1993

The tragic loss of the *Challenger* with its crew of seven astronauts on January 28, 1986, sent the space shuttle program into hiatus for 32 months. Seven astronauts died in this accident, then the worst in the history of spaceflight. The impact of the accident was felt far beyond NASA because schoolteacher Sharon Christa McAuliffe was aboard. She had been chosen from thousands of volunteers and was planning to teach a lesson from orbit. Accordingly, millions of schoolchildren tuned in to see her take off and watched as *Challenger* broke apart during launch. The shuttle did not return to action until September 29, 1988. During that time, NASA grappled with the technologically difficult issues associated with the *Challenger* accident.

After modifications to the shuttle's technical systems, management structures, and procedural priorities, the next launch of the space shuttle *Discovery* signaled a return to program operations. No doubt, the space shuttle was safer than ever before, and with a reinvigorated NASA under the leadership of former astronaut Vice Admiral Richard Truly, the program entered an era of exceptional accomplishment. Regardless, President Ronald Reagan directed that NASA retreat from providing launch services for other organizations, and removed commercial and Defense Department payloads from the vehicle.

During the period between return to flight in 1988 and the end of 1993, NASA flew 34 shuttle missions. As a workhorse, the space shuttle proved exceptional. Its missions delivered many significant scientific satellites to space, such as the Hubble Space Telescope in 1990. Its crews also demonstrated that they could rendezvous and service the Hubble Space Telescope, first doing so in December 1993.

1]
During the flight of STS-56, astronaut Ellen Ochoa plays a 15-minute set in *Discovery*'s aft flight deck. She performed the Marine Corps Hymn, Navy Hymn, and "God Save the Queen," as well as some Vivaldi.

2]
In the first servicing mission of the Hubble Space Telescope during STS-61 in December 1993, Story Musgrave (top right center) works to anchor the spacecraft for further repairs. Jeffrey Hoffman is also seen here working on the telescope. This was the first of five space walks necessary to service the telescope and ensure its optimum capability.

3]
During the same Hubble Space Telescope servicing mission in December 1993, astronaut Story Musgrave is barely visible in the payload bay (lower center) working with the Canadarm remote manipulator arm during the last EVA to service Hubble.

STS-26 (DISCOVERY)

LAUNCH DATE September 29, 1988
FLIGHT TIME 4 days, 1 hour
CREW Frederick "Rick" Hauck, Richard Covey, John "Mike" Lounge, David Hilmers, George Nelson

This first shuttle mission after the *Challenger* accident tested the safety of the redesigned solid rocket boosters. The *Discovery* crew were all veterans. They also tested launch and reentry space suits, and on orbit they practiced a new emergency escape system. The artwork for STS-26 depicts the launch of the space shuttle during its return to flight, with the sunburst in the background representing a glorious return to space orbital activities and the NASA red swoosh in the center reminiscent of the agency's logo. The Big Dipper to the left recalls the seven lost astronauts on *Challenger*'s last flight.

STS-27 (ATLANTIS)

LAUNCH DATE December 2, 1988
FLIGHT TIME 4 days, 9 hours, 6 minutes
CREW Robert "Hoot" Gibson, Guy Gardner, Richard Mullane, Jerry Ross, William Shepherd

This was a classified Department of Defense mission. This STS-27 mission insignia is reminiscent of earlier mission patches, with the multicolor background and shuttle launch. As in the insignia for STS-26, the seven stars remember the *Challenger* crew lost in 1986.

STS-29 (DISCOVERY)

LAUNCH DATE March 13, 1989
FLIGHT TIME 4 days, 9 hours, 39 minutes
CREW Michael Coats, John Blaha, Robert Springer, James Buchli, James Bagian

The crew released the fourth NASA tracking and data relay satellite into orbit and conducted several other scientific experiments. In a unique development, the astronauts also took large-format IMAX video footage of Earth. This mission patch emphasizes the shuttle with a sunburst behind it.

STS-30 (ATLANTIS)

LAUNCH DATE May 4, 1989
FLIGHT TIME 4 days, 56 minutes
CREW David Walker, Ronald Grabe, Mark Lee, Norman Thagard, Mary Cleave

This mission launched the first planetary probe from the payload bay, the *Magellan*. The unique emblem on the patch features the probe, showing the voyage from Earth's orbit to Venus as well as Ferdinand Magellan's circumnavigation of the globe.

STS-34 (ATLANTIS)

LAUNCH DATE October 18, 1989
FLIGHT TIME 4 days, 23 hours, 39 minutes
CREW Donald Williams, Michael McCulley, Shannon Lucid, Franklin Chang-Diaz, Ellen Baker

This mission deployed the Jupiter-bound *Galileo* spacecraft, which used gravity assist to reach the gas giant in 1995. This patch design also includes seven stars representing the *Challenger* crew.

STS-32 (COLUMBIA)

LAUNCH DATE January 9, 1990

FLIGHT TIME 10 days, 21 hours

CREW Daniel Brandenstein, James Wetherbee, Bonnie Dunbar, Marsha Ivins, David Low

This crew retrieved the long duration exposure facility that had been in orbit since 1984 and brought it back to Earth for analysis. The crew also released the *Syncom IV-5* satellite. In the process, this became the longest shuttle mission to that time. The design of this mission patch represents both achievements.

STS-31 (DISCOVERY)

LAUNCH DATE April 24, 1990

FLIGHT TIME 5 days, 1 hour, 16 minutes

CREW Loren Shriver, Charles Bolden, Bruce McCandless, Steven Hawley, Kathryn Sullivan

During this landmark mission, the STS-31 crew deployed the Hubble Space Telescope, the largest optical telescope ever placed in orbit. To achieve this, *Discovery* flew the highest shuttle orbit to date, reaching an altitude of more than 329.22 statute miles.

STS-41 (DISCOVERY)

LAUNCH DATE October 6, 1990

FLIGHT TIME 4 days, 2 hours, 10 minutes

CREW Richard Richards, Robert Cabana, Bruce Melnick, William Shepherd, Thomas Akers

This mission deployed the European Space Agency's *Ulysses* space probe that imaged the Sun's polar regions for the first time, and this is illustrated in the patch design.

STS-35 (COLUMBIA)

LAUNCH DATE December 2, 1990

FLIGHT TIME 8 days, 23 hours, 5 minutes

CREW Vance Brand, Guy Gardner, Jeffrey Hoffman, John "Mike" Lounge, Robert Parker, Samuel Durrance, Ronald Parise

The first *Spacelab* mission since the *Challenger* accident, STS-35 focused on astrophysics research. It carried the astronomical observatory ASTRO-1, which consisted of four telescopes. They were especially successful in observing X-ray and ultraviolet wavelengths.

STS-37 (ATLANTIS)

LAUNCH DATE April 5, 1991

FLIGHT TIME 5 days, 23 hours, 33 minutes

CREW Steven Nagel, Kenneth Cameron, Linda Godwin, Jerry Ross, Jerome "Jay" Apt

This mission deployed the Gamma Ray Observatory, following a space walk by Jerry Ross and Jay Apt to repair an antenna on the spacecraft. This was the second of NASA's four Great Observatories after the Hubble Space Telescope. The mission patch emphasizes its deployment.

STS-39 (DISCOVERY)

LAUNCH DATE April 28, 1991

FLIGHT TIME 8 days, 7 hours, 22 minutes

CREW Michael Coats, Blaine Hammond, Gregory Harbaugh, Donald McMonagle, Guion "Guy" Bluford, Charles Veach, Richard Hieb

This mission included extensive observations of space and returned high-quality images of Earth's aurora. Uniquely, *Discovery* released an instrument platform that flew in formation with the shuttle to observe rocket thruster plumes.

STS-40 (COLUMBIA)

LAUNCH DATE June 5, 1991

FLIGHT TIME 9 days, 2 hours, 14 minutes

CREW Bryan O'Connor, Sidney Gutierrez, James Bagian, Tamara Jernigan, Rhea Seddon, Andrew Gaffney, Millie Hughes-Fulford

This shuttle mission undertook Spacelab Life Sciences (SLS-1) experiments to understand physiological effects on the human body in space. Along with human subjects, rodents and jellyfish were also on board to be tested in microgravity.

STS-43 (ATLANTIS)

LAUNCH DATE August 2, 1991
FLIGHT TIME 8 days, 21 hours, 21 minutes
CREW John Blaha, Michael Baker, Shannon Lucid, David Low, James Adamson

With a smaller crew than normal, this mission deployed Tracking and Data Relay Satellite-5 (TDRS-5) and is largely remembered for the first landing at the Kennedy Space Center since the *Challenger* accident. This mission patch reflects the deployment of the TDRS-5.

STS-48 (DISCOVERY)

LAUNCH DATE September 12, 1991
FLIGHT TIME 5 days, 8 hours, 28 minutes
CREW John Creighton, Kenneth Reightler, Charles "Sam" Gemar, James Buchli, Mark Brown

The STS-48 crew deployed the Upper Atmosphere Research Satellite (UARS), a 14,400-pound observatory to investigate the stratosphere, mesosphere, and lower thermosphere. The STS-48 mission patch depicts the shuttle landing and includes the UARS experiment package.

STS-42 (DISCOVERY)

LAUNCH DATE January 22, 1992
FLIGHT TIME 8 days, 1 hour, 15 minutes
CREW Ronald Grabe, Stephen Oswald, Norman Thagard, William Readdy, David Hilmers, Roberta Bondar, Ulf Merbold

This marked the debut of the International Microgravity Laboratory (IML-1), a pressurized *Spacelab* module that investigated the human nervous system's adaptation to low gravity and the effects of microgravity on other life-forms. The STS-42 mission emblem emphasizes IML-1.

STS-45 (ATLANTIS)

LAUNCH DATE March 24, 1992
FLIGHT TIME 8 days, 22 hours, 9 minutes
CREW Charles Bolden, Brian Duffy, Kathryn Sullivan, David Leestma, Michael Foale, Dirk "Dick" Frimout, Byron Lichtenberg

The Atmospheric Laboratory for Applications and Science-1 debuted on this mission focusing on atmospheric sciences. The mission patch plays up ATLAS-1.

STS-49 (ENDEAVOUR)

LAUNCH DATE May 7, 1992
FLIGHT TIME 8 days, 21 hours, 18 minutes
CREW Daniel Brandenstein, Kevin Chilton, Richard Hieb, Bruce Melnick, Pierre Thuot, Kathryn Thornton, Thomas Akers

Two space walks on this mission were the longest in U.S. history to date, lasting eight hours, 29 minutes, and seven hours, 45 minutes each. This was also the first time three crew members worked outside the spacecraft at the same time.

STS-50 (COLUMBIA)

LAUNCH DATE June 25, 1992
FLIGHT TIME 13 days, 19 hours, 30 minutes
CREW Richard Richards, Kenneth Bowersox, Bonnie Dunbar, Ellen Baker, Carl Meade, Lawrence DeLucas, Eugene Trinh

The U.S. Microgravity Laboratory-1 made its first flight on STS-50. This was one of the longest-duration missions in U.S. spaceflight history, after the three *Skylab* missions in 1973 and 1974.

STS-46 (ATLANTIS)

LAUNCH DATE July 31, 1992
FLIGHT TIME 7 days, 23 hours, 15 minutes
CREW Loren Shriver, Andrew Allen, Claude Nicollier, Marsha Ivins, Jeffrey Hoffman, Franklin Chang-Díaz, Franco Malerba

This mission deployed the European Space Agency's European Retrievable Carrier and tested the NASA/Italian Space Agency tethered satellite system. The crew reeled it in for a test on a later mission.

STS-47 (ENDEAVOUR)

LAUNCH DATE September 12, 1992
FLIGHT TIME 7 days, 22 hours, 30 minutes
CREW Robert "Hoot" Gibson, Curtis Brown, Mark Lee, Jerome "Jay" Apt, Jan Davis, Mae Jemison, Mamoru Mohri

The STS-47 mission featured *Spacelab-J*, the first Japanese space laboratory, containing 24 experiments in materials science, and 20 in life science in a range of disciplines. The Japanese and American flags are both represented in the mission patch.

STS-52 (COLUMBIA)

LAUNCH DATE October 22, 1992
FLIGHT TIME 9 days, 20 hours, 56 minutes
CREW James Wetherbee, Michael Baker, Charles Veach, William Shepherd, Tamara Jernigan, Steven MacLean

STS-52 deployed the Laser Geodynamic Satellite II (LAGEOS), a joint effort of NASA and the Italian Space Agency, and operated the U.S. Microgravity Payload-1 (USMP-1). This mission patch shows the shuttle in orbit along with LAGEOS.

STS-54 (ENDEAVOUR)

LAUNCH DATE January 13, 1993
FLIGHT TIME 5 days, 23 hours, 38 minutes
CREW John Casper, Donald McMonagle, Mario Runco, Gregory Harbaugh, Susan Helms

STS-54 deployed the fifth tracking and data relay satellite as part of NASA's orbiting communications system. The mission patch design has more in common with the *Apollo 11* patch than any other.

STS-56 (DISCOVERY)

LAUNCH DATE April 8, 1993

FLIGHT TIME 9 days, 6 hours, 8 minutes

CREW Kenneth Cameron, Stephen Oswald, Michael Foale, Kenneth Cockrell, Ellen Ochoa

This mission featured the Atmospheric Laboratory for Applications and Science-2 (ATLAS-2), which collected data on the Sun's energy output in relation to Earth's atmosphere. One of the more creative mission patches, this depicts only a portion of the space shuttle, emphasizing the work of ATLAS-2 in the payload bay.

STS-55 (COLUMBIA)

LAUNCH DATE April 26, 1993

FLIGHT TIME 9 days, 23 hours, 40 minutes

CREW Steven Nagel, Terence Henricks, Jerry Ross, Charles Precourt, Bernard Harris, Ulrich Walter, Hans Schlegel

For the second time, NASA and the German Agency for Space Flight Affairs teamed up to fly a *Spacelab* mission, involving experiments in materials and life sciences, Earth observations, and astronomy. The patch depicts this partnership with both flags.

STS-57 (ENDEAVOUR)

LAUNCH DATE June 21, 1993
FLIGHT TIME 9 days, 23 hours, 45 minutes
CREW Ronald Grabe, Brian Duffy, David Low, Nancy Sherlock, Peter Wisoff, Janice Voss

The SPACEHAB module flown in the payload bay of STS-57 took center stage on this patch. Its comprehensive experiments package supported materials and life sciences studies around the globe. The crew also deployed the European Retrievable Carrier with a long-duration experiments package.

STS-51 (DISCOVERY)

LAUNCH DATE September 12, 1993
FLIGHT TIME 9 days, 20 hours, 11 minutes
CREW Frank Culbertson, William Readdy, James Newman, Daniel Bursch, Carl Walz

This mission deployed the Advanced Communications Technology Satellite (ACTS) to geosynchronous transfer orbit. It also deployed a shuttle pallet satellite to investigate celestial sources and increase human understanding of the evolution of stars. It was the first in a series of German–U.S. astronomical missions.

STS-58 (COLUMBIA)

LAUNCH DATE October 18, 1993
FLIGHT TIME 14 days, 13 minutes
CREW John Blaha, Richard Searfoss, Rhea Seddon, William McArthur, David Wolf, Shannon Lucid, Martin Fettman

The STS-58 crew undertook the second dedicated Spacelab Life Sciences mission and carried out 14 experiments. Eight centered on crew health, while six focused on 48 rodents carried on board. Complete with a border reminiscent of the symbol of medicine, this is an unusual and attractive design for the STS-58 mission patch.

STS-61 (ENDEAVOUR)

LAUNCH DATE December 2, 1993
FLIGHT TIME 10 days, 19 hours, 58 minutes
CREW Richard Covey, Kenneth Bowersox, Kathryn Thornton, Claude Nicollier, Jeffrey Hoffman, Story Musgrave, Thomas Akers

In one of the most rigorous and anticipated missions in shuttle history, the STS-61 crew returned to the Hubble Space Telescope to service it after anomalies were found since its deployment in 1990. The crew undertook a record five back-to-back space walks totaling 35 hours and 28 minutes. A truly memorable mission, the insignia reflects the crew's preference for a simple geometric design. The space shuttle is small, while the orbit and the Hubble-servicing aspect of the mission is represented by joined circles. The astronaut symbol is at the center of the design.

CLASSIFIED MISSIONS

Even before the space shuttle began flying in the 1970s, the U.S. Department of Defense negotiated for it to be used for selected national security missions of a classified nature. Several of these missions were flown during the 1980s. There is little information about these flights and the record includes few details other than launch and landing information, although many have speculated that they deployed national security—possibly reconnaissance—satellites into orbit.

STS-28 (COLUMBIA)

LAUNCH DATE August 8, 1989
FLIGHT TIME 5 days, 1 hour
CREW Brewster Shaw, Richard Richards, James Adamson, David Leestma, Mark Brown

STS-36 (ATLANTIS)

LAUNCH DATE February 28, 1990
FLIGHT TIME 4 days, 10 hours, 18 minutes
CREW John Creighton, John Casper, Richard Mullane, David Hilmers, Pierre Thuot

STS-33 (DISCOVERY)

LAUNCH DATE November 22, 1989
FLIGHT TIME 5 days, 7 minutes
CREW Frederick Gregory, John Blaha, Manley "Sonny" Carter, Story Musgrave, Kathryn Thornton

STS-38 (ATLANTIS)

LAUNCH DATE November 15, 1990
FLIGHT TIME 4 days, 21 hours, 55 minutes
CREW Richard Covey, Frank Culbertson, Robert Springer, Carl Meade, Charles "Sam" Gemar

STS-44 (ATLANTIS)

LAUNCH DATE November 24, 1991
FLIGHT TIME 6 days, 22 hours, 51 minutes
CREW Frederick Gregory, Terence Henricks, James Voss, Story Musgrave, Mario Runco, Thomas Hennen

STS-53 (DISCOVERY)

LAUNCH DATE December 2, 1992
FLIGHT TIME 7 days, 7 hours, 20 minutes
CREW David Walker, Robert Cabana, Guion "Guy" Bluford, James Voss, Michael Clifford

CHAPTER 7
SPACE SHUTTLE
1994–1998

During the early 1990s, NASA and the Russian Aviation and Space Agency negotiated a cooperative shuttle/*Mir* program to undertake joint space missions. Among other aspects of the program, the space shuttle undertook several visits to the Russian space station *Mir* in the middle part of the decade. During the third visit in March 1996, crew members aboard *Mir* snapped this picture of the space shuttle *Atlantis* with the Indian Ocean in the background. The solar arrays in the foreground are those supplying power to *Mir*.

SPACE SHUTTLE 1994–1998

In the aftermath of the Cold War, NASA began negotiations with the Russian Aviation and Space Agency for a cooperative human spaceflight program. The centerpiece of this program was a set of docking missions in Earth's orbit during the mid-1990s. The Russian *Mir* space station had been launched in 1986, and between 1995 and 1998 a series of nine joint shuttle/*Mir* missions unfolded. Kicking off the joint effort, in February 1995 Russia's most experienced cosmonaut, Sergei Krikalev, flew as part of STS-60. Equally significant, in March, American astronaut Norman Thagard flew to *Mir* on the Russian *Soyuz-TM 21* spacecraft. He remained there until the summer of 1995.

During STS-71, the space shuttle *Atlantis* docked to the *Mir* space station in an historic transfer of crew and cargo. After commemorative events aboard *Mir*, the two groups of spacefarers undertook several days of joint scientific investigations inside the *Spacelab* module tucked into *Atlantis*'s large cargo bay. Research in several medical and scientific disciplines begun on *Mir* also concluded on STS-71. At the end of joint docked activities, cosmonauts Anatoly Solovyev and Nikolai Budarin assumed responsibility for operations on *Mir* while Thagard and other crew members returned to Earth on the shuttle.

During this first *Mir* docking mission, as well as those that followed, the Americans and Russians learned to cooperate effectively, with the space shuttle program leading the way. They gained experience working together in orbit, no doubt, but sometimes the road was rocky. Both nations had a long history of leadership in space, and they were unaccustomed to negotiation and compromise. The logical partnership, however, changed both organizations for the better. In the process, during the mid-1990s the shuttle program planned and developed the capabilities necessary to undertake its final and most significant task, the building of a major space station in orbit.

Even as these joint missions progressed, other shuttle flights continued to take place. In each case, the shuttle crews designed their own mission patches, reflecting the major elements of the flight and the names of the crew. Each became an individualized statement of crew activity.

1]
Taken from a Soyuz spacecraft, the space shuttle *Atlantis* is shown docked to the *Mir* space station in June 1995. This was the first of nine missions that docked with the Russian space station.

2]
U.S. senator John Glenn had lobbied NASA for years to allow him to return to space. In October 1998, he flew as a payload specialist on STS-95 researching the effects of aging on the body.

STS-60 (DISCOVERY)

LAUNCH DATE February 3, 1994
FLIGHT TIME 8 days, 7 hours, 9 minutes
CREW Charles Bolden, Kenneth Reightler, Jan Davis, Ronald Sega, Franklin Chang-Diaz, Sergei Krikalev

STS-60 marked the first flight of a Russian cosmonaut on a U.S. space shuttle, inaugurating the shuttle/Mir joint program. The patch artwork depicts the Canadarm deploying a satellite, superimposed over the flags of the two countries.

STS-62 (COLUMBIA)

LAUNCH DATE March 4, 1994
FLIGHT TIME 13 days, 23 hours, 17 minutes
CREW John Casper, Andrew Allen, Pierre Thuot, Charles "Sam" Gemar, Marsha Ivins

The U.S. Microgravity Payload-2 and the Office of Aeronautics and Space Technology-2 were the focus of this mission. The microgravity experiments investigated materials processing and crystal growth in microgravity. OAST-2 focused on space technology and spaceflight.

STS-59 (ENDEAVOUR)

LAUNCH DATE April 9, 1994
FLIGHT TIME 11 days, 5 hours, 50 minutes
CREW Sidney Gutierrez, Kevin Chilton, Jerome "Jay" Apt, Michael Clifford, Linda Godwin, Thomas Jones

The Space Radar Laboratory-1 located in the payload bay gathered data about Earth and the effect of human activity on its carbon, water, and energy cycles. It was a joint project of NASA, the German Agency for Space Flight Affairs, and the Italian Space Agency.

STS-65 (COLUMBIA)

LAUNCH DATE July 8, 1994

FLIGHT TIME 14 days, 17 hours, 55 minutes

CREW Robert Cabana, James Halsell, Richard Hieb, Carl Walz, Leroy Chiao, Donald Thomas, Chiaki Naito-Mukai

This flight included the first Japanese woman to fly in space, payload specialist Chiaki Naito-Mukai. The primary payload was the International Microgravity Laboratory, which flew for the second time. Crew members tended more than 80 experiments in this laboratory, working with more than 200 scientists around the globe. This flight was the longest shuttle flight to date. The swoosh of the shuttle around the second International Microgravity Laboratory gives this mission patch a distinctive style.

STS-64 (DISCOVERY)

LAUNCH DATE September 9, 1994

FLIGHT TIME 10 days, 22 hours, 50 minutes

CREW Richard Richards, Blaine Hammond, Jerry Linenger, Susan Helms, Carl Meade, Mark Lee

This mission featured the Lidar In-Space Technology Experiment, using lasers for environmental research. It also included the first untethered U.S. extravehicular activity in a decade. The crew released and retrieved the SPARTAN-201 using the remote manipulator system arm. This unique patch features the astronaut emblem at its center, with astronauts in untethered EVAs on either side and the space shuttle overhead. You can also see the payload bay doors and the Canadarm remote manipulator being deployed.

STS-68 (ENDEAVOUR)

LAUNCH DATE September 30, 1994
FLIGHT TIME 11 days, 5 hours, 46 minutes
CREW Michael Baker, Terrence Wilcutt, Steven Smith, Daniel Bursch, Peter Wisoff, Thomas Jones

The Space Radar Laboratory flew for the second time, repeating earlier observations of Earth. A major feature of this effort involved imaging radar that distinguished between changes caused by human-induced phenomena, such as oil spills, and naturally occurring events.

STS-66 (ATLANTIS)

LAUNCH DATE November 3, 1994
FLIGHT TIME 10 days, 22 hours, 34 minutes
CREW Donald McMonagle, Curtis Brown, Ellen Ochoa, Joseph Tanner, Jean-François Clervoy, Scott Parazynski

This mission emphasized research using the Atmospheric Laboratory for Applications and Science-3 experiments module in the payload bay, which was featured on the mission patch.

STS-63 (DISCOVERY)

LAUNCH DATE February 3, 1995
FLIGHT TIME 8 days, 6 hours, 28 minutes
CREW James Wetherbee, Eileen Collins, Bernard Harris, Michael Foale, Janice Voss, Vladimir Titov

This was the first of the joint missions flown by NASA with the Roscosmos State Corporation for Space Activities. *Discovery* rendezvoused with the Russian *Mir* space station. The mission patch commemorates this historic meeting in Earth's orbit by depicting the space shuttle and a portion of *Mir*.

STS-67 (ENDEAVOUR)

LAUNCH DATE March 2, 1995

FLIGHT TIME 16 days, 15 hours, 9 minutes

CREW Stephen Oswald, William Gregory, John Grunsfeld, Wendy Lawrence, Tamara Jernigan, Samuel Durrance, Ronald Parise

STS-67 set an endurance record in completing the longest shuttle flight to date. Its primary scientific payload was ASTRO-2, a suite of three ultraviolet telescopes. The mission patch depicts several cosmological phenomena measured by ASTRO-2.

STS-71 (ATLANTIS)

LAUNCH DATE June 27, 1995

FLIGHT TIME 9 days, 19 hours, 22 minutes

CREW Robert Gibson, Charles Precourt, Ellen Baker, Gregory Harbaugh, Bonnie Dunbar, Norman Thagard, Anatoly Solovyev, Nikolai Budarin, Vladimir Dezhurov, Gennady Strekalov

The 100th U.S. human spaceflight was memorable because it inaugurated a series of docking missions with the Russian space station *Mir*, depicted by the interlocking spheres on the patch.

STS-70 (DISCOVERY)

LAUNCH DATE July 13, 1995

FLIGHT TIME 8 days, 22 hours, 20 minutes

CREW Terence Henricks, Kevin Kregel, Nancy Currie, Donald Thomas, Mary Weber

In addition to the deployment of Tracking and Data Relay Satellite-G used in NASA's global space communications system, STS-70 also conducted experiments on the effects of microgravity on several organisms. The patch design with a shuttle flying over Earth suggests the link between space-based research and its applications on Earth.

STS-69 (ENDEAVOUR)

LAUNCH DATE September 7, 1995
FLIGHT TIME 10 days, 20 hours, 29 minutes
CREW David Walker, Kenneth Cockrell, James Voss, James Newman, Michael Gernhardt

STS-69's primary objective was the deployment and retrieval of a satellite measuring aspects of the outer atmosphere of the Sun. It also deployed and retrieved the Wake Shield Facility-2, which generated an ultravacuum environment for growing thin films used in semiconductors.

STS-73 (COLUMBIA)

LAUNCH DATE October 20, 1995
FLIGHT TIME 15 days, 21 hours, 52 minutes
CREW Kenneth Bowersox, Kent Rominger, Kathryn Thornton, Catherine Coleman, Michael Lopez-Alegria, Fred Leslie, Albert Sacco

A second flight of the United States Microgravity Laboratory repeated some of the same experiments from the first USML mission from STS-50 to replicate results. This STS-73 mission patch looks similar to the STS-29 emblem (see page 103).

STS-74 (ATLANTIS)

LAUNCH DATE November 12, 1995
FLIGHT TIME 8 days, 4 hours, 31 minutes
CREW Kenneth Cameron, James Halsell, Chris Hadfield, Jerry Ross, William McArthur

STS-74 is remembered largely for the second in a series of nine *Mir* space station dockings. The mission patch reflects this, showing the shuttle and the Russian Docking Module attached to *Mir*, which it also delivered.

STS-72 (ENDEAVOUR)

LAUNCH DATE January 11, 1996
FLIGHT TIME 8 days, 22 hours, 2 minutes
CREW Brian Duffy, Brent Jett, Leroy Chiao, Winston Scott, Daniel Barry, Koichi Wakata

STS-72 focused on the capture and return to Earth of the *Space Flyer Unit*, a Japanese microgravity research spacecraft that had been in space since March 1995. The mission patch depicted the space shuttle with the unit in a free-flying configuration above the payload bay.

STS-75 (COLUMBIA)

LAUNCH DATE February 22, 1996
FLIGHT TIME 15 days, 17 hours, 41 minutes
CREW Andrew Allen, Scott Horowitz, Franklin Chang-Diaz, Jeffrey Hoffman, Maurizio Cheli, Claude Nicollier, Umberto Guidoni

This mission resumed an experiment to generate usable electricity by dangling receptors through charged portions of Earth's atmosphere, but the tether broke three days into the mission.

STS-76 (ATLANTIS)

LAUNCH DATE March 22, 1996
FLIGHT TIME 9 days, 5 hours, 16 minutes
CREW Kevin Chilton, Richard Searfoss, Ronald Sega, Michael Clifford, Linda Godwin, Shannon Lucid

The third docking mission with the Russian space station *Mir*, STS-76 also included a space walk, logistics operations, and scientific research. Nearly 1,900 pounds of equipment was transferred from *Atlantis* to *Mir*. The STS-76 mission patch features *Mir* and the space shuttle.

STS-77 (ENDEAVOUR)

LAUNCH DATE May 19, 1996
FLIGHT TIME 10 days, 39 minutes
CREW John Casper, Curtis Brown, Andrew Thomas, Daniel Bursch, Mario Runco, Marc Garneau

The crew performed microgravity research aboard the commercial Spacehab module, and deployed and retrieved the Spartan-207 Inflatable Antenna Experiment satellite. The mission patch uniquely shows two versions of the orbiter.

STS-78 (COLUMBIA)

LAUNCH DATE June 20, 1996
FLIGHT TIME 16 days, 21 hours, 48 minutes
CREW Terence Henricks, Kevin Kregel, Richard Linnehan, Susan Helms, Charles Brady, Jean-Jacques Favier, Robert Thirsk

The Life and Microgravity Spacelab module in the payload bay facilitated a range of experiments on human physiology and space biology. The most interesting element of the mission patch was the inclusion of Native American symbology.

STS-79 (ATLANTIS)

LAUNCH DATE September 16, 1996
FLIGHT TIME 10 days, 3 hours, 19 minutes
CREW William Readdy, Terrence Wilcutt, Jerome "Jay" Apt, Thomas Akers, Carl Walz, John Blaha, Shannon Lucid

This mission saw the fourth shuttle docking with *Mir*. Shannon Lucid also set a U.S. women's record for length of time in space: 188 days and five hours. When she returned to Earth, John Blaha replaced her on *Mir*. The mission patch depicts a handclasp between Russian and American crew members.

STS-80 (COLUMBIA)

LAUNCH DATE November 19, 1996
FLIGHT TIME 17 days, 15 hours, 53 minutes
CREW Kenneth Cockrell, Kent Rominger, Tamara Jernigan, Thomas Jones, Story Musgrave

This mission deployed and retrieved the Wake Shield Facility and the Orbiting and Retrievable Far and Extreme Ultraviolet Spectrometer–Shuttle Pallet Satellite II. Both are depicted on the mission patch above and below the curved red lines suggesting the astronaut emblem.

STS-81 (ATLANTIS)

LAUNCH DATE January 12, 1997
FLIGHT TIME 10 days, 4 hours, 55 minutes
CREW Michael Baker, Brent Jett, Peter Wisoff, John Grunsfeld, Marsha Ivins, Jerry Linenger, John Blaha

A fifth *Mir* docking mission, STS-81 exchanged astronaut Jerry Linenger for John Blaha after his 128 days in space. *Atlantis* also carried a Spacehab double module, which provided additional space for human-tended experiments.

STS-82 (DISCOVERY)

LAUNCH DATE February 11, 1997
FLIGHT TIME 9 days, 23 hours, 37 minutes
CREW Kenneth Bowersox, Scott Horowitz, Joseph Tanner, Steven Hawley, Gregory Harbaugh, Mark Lee, Steven Smith

STS-82 undertook the second servicing mission of the Hubble Space Telescope, making upgrades to extend its service life. The STS-82 mission patch depicts only the Hubble in orbit.

STS-83 (COLUMBIA)

LAUNCH DATE April 4, 1997
FLIGHT TIME 3 days, 23 hours, 14 minutes
CREW James Halsell, Susan Still, Janice Voss, Michael Gernhardt, Donald Thomas, Roger Crouch, Gregory Linteris

This mission was cut short by a failure of one of the fuel cells, meaning the experimentation on the Microgravity Science Laboratory-1 had to be rescheduled for later. The mission patch emphasizes the MSL aboard the shuttle.

STS-84 (ATLANTIS)

LAUNCH DATE May 15, 1997
FLIGHT TIME 9 days, 23 hours, 20 minutes
CREW Charles Precourt, Eileen Collins, Jean-François Clervoy, Carlos Noriega, Edward Lu, Yelena Kondakova, Michael Foale, Jerry Linenger

This sixth shuttle/*Mir* docking mission took U.S. astronaut Mike Foale aboard *Mir* and brought Jerry Linenger back. Depicting the space shuttle at liftoff, this mission patch also shows *Mir* in Cyrillic.

STS-94 (COLUMBIA)

LAUNCH DATE July 1, 1997
FLIGHT TIME 15 days, 16 hours, 44 minutes
CREW James Halsell, Susan Still, Janice Voss, Michael Gernhardt, Donald Thomas, Roger Crouch, Gregory Linteris

This was the reflight of STS-83, which had the Microgravity Science Laboratory aboard. It investigated the routine influence of gravity on materials and liquids. This mission patch was a close rendition of what had been created for STS-83 (see above).

STS-85 (DISCOVERY)

LAUNCH DATE August 7, 1997
FLIGHT TIME 11 days, 19 hours, 19 minutes
CREW Curtis Brown, Kent Rominger, Robert Curbeam, Jan Davis, Stephen Robinson, Bjarni Tryggvason

Focusing on Earth science, STS-85 carried the Cryogenic Infrared Spectrometers and Telescopes for the Atmosphere–Shuttle Pallet Satellite-2, the Technology Applications and Science-01, and the International Extreme Ultraviolet Hitchhiker-02. A busy mission patch seeks to represent each of these elements, as well as a new Japanese robotic arm designed for use on a future space station.

STS-86 (ATLANTIS)

LAUNCH DATE September 25, 1997
FLIGHT TIME 10 days, 19 hours, 21 minutes
CREW James Wetherbee, Michael Bloomfield, Vladimir Titov, Scott Parazynski, Jean-Loup Chrétien, Wendy Lawrence, David Wolf, Michael Foale

This seventh *Mir* docking mission transferred American astronaut David Wolf to *Mir* while returning Mike Foale to Earth. The mission patch plays up the international collaboration on *Mir*, depicting the crew in their native language, using the colors of the American and Russian flags, and showing EVAs by each nationality. Wolf's name is featured outside the patch proper because he was flying to *Mir* and was not a member of the crew.

STS-87 (COLUMBIA)

LAUNCH DATE November 19, 1997
FLIGHT TIME 15 days, 16 hours, 34 minutes
CREW Kevin Kregel, Steven Lindsey, Kalpana Chawla, Winston Scott, Takao Doi, Leonid Kadenyuk

The mission patch reflects the emphasis of STS-87 on microgravity science with the "µg," the symbol for microgravity. Uniquely, the flag of Ukraine recognizes Leonid Kadenyuk, the first Ukrainian to fly on the space shuttle.

STS-89 (ENDEAVOUR)

LAUNCH DATE January 22, 1998
FLIGHT TIME 8 days, 19 hours, 47 minutes
CREW Terrence Wilcutt, Joe Edwards, James Reilly, Michael Anderson, Bonnie Dunbar, Salizhan Sharipov, Andrew Thomas, David Wolf

The eighth *Mir* shuttle docking featured the exchange of David Wolf for Andrew Thomas. This joint U.S.–Russian mission is reflected in the patch with its depiction of the shuttle docked to *Mir*.

STS-90 (COLUMBIA)

LAUNCH DATE April 17, 1998
FLIGHT TIME 15 days, 21 hours, 51 minutes
CREW Richard Searfoss, Scott Altman, Richard Linnehan, Kathryn Hire, Dafydd Williams, Jay Buckey, James Pawelczyk

This dedicated Neurolab flight, with a version of the longstanding *Spacelab* module affixed to the payload bay, allowed 26 experiments relating to the human nervous system. A cooperative venture between many national space agencies, the mission patch showed the universe and the shuttle in orbit.

STS-91 (DISCOVERY)

LAUNCH DATE June 2, 1998
FLIGHT TIME 9 days, 19 hours, 54 minutes
CREW Charles Precourt, Dominic Gorie, Franklin Chang-Diaz, Wendy Lawrence, Janet Kavandi, Valery Ryumin, Andrew Thomas

STS-91 marked the final *Mir* docking and the end of phase one of Russian space collaboration. The next phase would involve building the International Space Station (ISS). The mission patch pays tribute to this program, depicting both the shuttle and *Mir*.

STS-95 (DISCOVERY)

LAUNCH DATE October 29, 1998
FLIGHT TIME 8 days, 21 hours, 44 minutes
CREW Curtis Brown, Steven Lindsey, Stephen Robinson, Scott Parazynski, Pedro Duque, Chiaki Naito-Mukai, John Glenn

This mission saw John Glenn return to space, evoking the early excitement of spaceflight from Project Mercury. With the symbol of the Mercury Seven astronauts, the mission patch also emphasizes the three scientific fields of the mission.

STS-88 (ENDEAVOUR)

LAUNCH DATE December 4, 1998
FLIGHT TIME 11 days, 19 hours, 18 minutes
CREW Robert Cabana, Frederick Sturckow, Jerry Ross, Nancy Currie, James Newman, Sergei Krikalev

STS-88 marked the first U.S. International Space Station assembly flight. As reflected in the mission patch, the Unity module was deployed and connected to the Zarya control module. Three space walks by Jerry Ross and James Newman ensured connectivity between Zarya and Unity.

A beautifully lit space shuttle *Atlantis* at dawn is reflected in rain puddles at Launch Complex 39A on July 7, 2011. The last launch of a space shuttle took place on that date, the 135th mission of the venerable spacecraft.

CHAPTER 8
SPACE SHUTTLE
1999–2011

SPACE SHUTTLE 1999–2011

From the point at which NASA began constructing the International Space Station (ISS) at the end of 1998 until the termination of the space shuttle program with the flight of STS-135 in 2011, NASA concentrated its human spaceflight program on two major objectives. First, its space shuttle missions undertook the construction, resupply, and crew rotation of the ISS. Second, the space shuttle undertook three additional Hubble Space Telescope servicing missions to ensure that this astronomical observatory continued its groundbreaking research into the 2020s.

NASA leaders were remarkably successful in maintaining an international political coalition of 15 nations supporting the effort. Assembly began in December 1998 with the flight of STS-88 and continued until the tragic *Columbia* accident of February 1, 2003, when seven astronauts lost their lives, and then continued in 2005 with the shuttle's return to flight. NASA declared the ISS complete in 2011. The first crew went aboard in the fall of 2000, and a total of 63 crews have served aboard this station through January 2020.

The space shuttle's retirement in 2011 sparked an assessment of the program. Over 30 years of flight from 1981 to 2011, the shuttle had proven itself a venerable space vehicle. Three legacies stood out. First, the space shuttle was a triumph of engineering and excellence in technological management. Second, it proved itself one of the most flexible space vehicles ever flown. With its large payload bay it could deploy, capture, and return satellites to Earth. No flights demonstrated the flexibility of the space shuttle more effectively than the Hubble Space Telescope servicing missions carried out between 1993 and 2010. Finally, the shuttle served as a marvelous platform for scientific inquiry. Each of its flights undertook scientific experiments, ranging from the deployment of important space probes to other planets, through the periodic use of the European-built *Spacelab* and other science modules, to a dramatic set of Earth observations over a 30-year period. The vehicle would be missed.

1] During the flight of STS-115 in September 2006, astronaut Daniel Burbank works on the assembly of the International Space Station in Earth's orbit.

2] The crew of STS-88 began construction of the International Space Station in orbit in December 1998 with the mating of the American Unity module with the Russian Zarya component. Here astronaut James Newman pauses during an EVA to wave to the camera with the new station components in the background.

STS-96 (DISCOVERY)

LAUNCH DATE May 27, 1999

FLIGHT TIME 9 days, 19 hours, 14 minutes

CREW Kent Rominger, Rick Husband, Tamara Jernigan, Ellen Ochoa, Daniel Barry, Julie Payette, Valery Tokarev

In this lengthy mission, *Discovery* docked with the ISS. Construction ensued, with a space walk to install both U.S.- and Russian-built cranes. The artwork for the badge of STS-96 uses a red, white, and blue astronaut emblem of three prongs reaching an apex as its central component. It also depicts the space shuttle during orbital operations, rendezvousing with the ISS in the background.

STS-93 (COLUMBIA)

LAUNCH DATE July 23, 1999

FLIGHT TIME 4 days, 22 hours, 50 minutes

CREW Eileen Collins, Jeffrey Ashby, Catherine Coleman, Steven Hawley, Michel Tognini

The objective of this mission, the shortest since 1990, was the deployment of the Chandra X-ray Observatory. This was the third of NASA's Great Observatories, following the Hubble Space Telescope and the Compton Gamma Ray Observatory. The insignia on the badge of the STS-93 mission features the deployment of the observatory and astronautical objects in the background. It also emphasizes the cooperative effort between the U.S. and France, with flags of each nation depicted.

STS-103 (DISCOVERY)

LAUNCH DATE December 19, 1999
FLIGHT TIME 7 days, 23 hours, 11 minutes
CREW Curtis Brown, Scott Kelly, Steven Smith, Jean-François Clervoy, John Grunsfeld, Michael Foale, Claude Nicollier

This was the third Hubble Space Telescope servicing mission. After the telescope's capture by the robotic arm, it was placed on the Flight Support System in *Discovery*'s cargo bay. Repairs and upgrades required three separate space walks. With the names of the crew surrounding the graphic, the STS-103 mission emblem depicts the space shuttle in orbit and a prominent Hubble Space Telescope, which its crew serviced.

STS-99 (ENDEAVOUR)

LAUNCH DATE February 11, 2000
FLIGHT TIME 11 days, 5 hours, 39 minutes
CREW Kevin Kregel, Dominic Gorie, Gerhard Thiele, Janet Kavandi, Janice Voss, Mamoru Mohri

During this mission, the STS-99 crew deployed the Shuttle Radar Topography Mission (SRTM) mast for mapping the Earth in stereo. The mission insignia depicted the SRTM imaging, featuring both stereoscopic cameras, one in the shuttle payload bay, and the other on a boom extending away from the spacecraft. The synchronized imagery enabled striking visualizations of the Earth's surface.

STS-101 (ATLANTIS)

LAUNCH DATE May 19, 2000
FLIGHT TIME 9 days, 20 hours, 9 minutes
CREW James Halsell, Scott Horowitz, Mary Weber, Jeffrey Williams, James Voss, Susan Helms, Yury Usachev

The third ISS construction flight, STS-101, installed the final parts of the Russian-built Strela crane, and replaced a faulty antenna. After docking with the ISS, the crew worked on outfitting the interior and preparing for the arrival of the Zvezda service module. The colorful and patriotic oval emblem on the mission patch emphasizes the construction of the ISS on orbit.

STS-106 (ATLANTIS)

LAUNCH DATE September 8, 2000
FLIGHT TIME 11 days, 19 hours, 12 minutes
CREW Terrence Wilcutt, Scott Altman, Edward Lu, Richard Mastracchio, Daniel Burbank, Yuri Malenchenko, Boris Morukov

During a space walk by Edward Lu and Yuri Malenchenko, the crew routed and connected power, data, and communications cables between the Zvezda and Zarya modules. It also transferred tons of supplies and equipment from *Atlantis* to the ISS. A slightly elongated oval-shaped emblem, the mission patch emphasizes the construction of the ISS as several elements are joined into one, as well as the cooperation between Russia and the U.S.

STS-92 (DISCOVERY)

LAUNCH DATE October 11, 2000

FLIGHT TIME 12 days, 21 hours, 43 minutes

CREW Brian Duffy, Pamela Melroy, Leroy Chiao, William McArthur, Peter Wisoff, Michael Lopez-Alegria, Koichi Wakata

In this fifth ISS construction flight, the crew installed the Zenith 1 truss and the third Pressurized Mating Adapter. The crew also transferred equipment and supplies to support the first resident crew arriving near the end of the year. The new installations are both reflected in the mission emblem with the "Z" swoosh making up part of the astronaut emblem and the adapter at the bottom of the ISS silhouette.

STS-97 (ENDEAVOUR)

LAUNCH DATE November 30, 2000

FLIGHT TIME 10 days, 19 hours, 58 minutes

CREW Brent Jett, Michael Bloomfield, Joseph Tanner, Marc Garneau, Carlos Noriega

This was the sixth ISS construction flight. It involved three space walks to accomplish the mating of the Port 6 truss and the first set of U.S.-provided solar arrays to the station's Zenith 1 truss. The construction flight is highlighted in this mission patch; this time the shuttle is docked to the space station over a colorful Earth below.

STS-98 (ATLANTIS)

LAUNCH DATE February 7, 2001

FLIGHT TIME 12 days, 20 hours, 20 minutes

CREW Kenneth Cockrell, Mark Polansky, Robert Curbeam, Marsha Ivins, Thomas Jones

In the seventh ISS construction flight, STS-98 delivered, attached, and activated the Destiny laboratory module to the station. This required three space walks lasting 19 hours and 49 minutes. One of the most unique shuttle mission patches, STS-98 is depicted in orbit with the Destiny laboratory module being affixed to the ISS with the Canadarm.

STS-102 (DISCOVERY)

LAUNCH DATE March 8, 2001

FLIGHT TIME 12 days, 19 hours, 49 minutes

CREW James Wetherbee, James Kelly, Andrew Thomas, Paul Richards

STS-102 was the eighth ISS construction flight. It also swapped out the Expedition 1 with Expedition 2 ISS crews. The Expedition 2 crew names were also included on the bottom of the patch in English and Russian. The patch also swirls the colors of the U.S. and Russian flags around the nascent ISS being constructed.

STS-100 (ENDEAVOUR)

LAUNCH DATE April 19, 2001

FLIGHT TIME 11 days, 12 hours, 54 minutes

CREW Kent Rominger, Jeffrey Ashby, Chris Hadfield, John Phillips, Scott Parazynski, Umberto Guidoni, Yuri Lonchakov

STS-100 carried the Canadian-built Canadarm 2, which the crew attached outside the station's Destiny module. The crew also delivered the second Multi-Purpose Logistics Module, named Raffaello. An elegant design, the STS-100 mission patch uses a striking color palette to create a memorable symbol of an ISS construction flight. The patch itself is shaped like a space suit helmet.

STS-104 (ATLANTIS)

LAUNCH DATE July 12, 2001

FLIGHT TIME 12 days, 18 hours, 36 minutes

CREW Steven Lindsey, Charles Hobaugh, Michael Gernhardt, Janet Kavandi, James Reilly

The 10th ISS construction flight delivered the Quest Airlock, which allowed for space walks from the station. During space walks to accomplish mission requirements, astronauts spent 16 hours and 30 minutes outside the station. With the astronaut symbol in the center and the shuttle in orbit, the mission patch features the American flag. The overall shape of the patch is the same as the Pentagon, the home of the U.S. Department of Defense.

STS-105 (DISCOVERY)

LAUNCH DATE August 10, 2001
FLIGHT TIME 11 days, 19 hours, 38 minutes
CREW Scott Horowitz, Frederick Sturckow, Patrick Forrester, Daniel Barry

Discovery performed the 11th ISS construction flight, replacing crews on the station and bringing up a load of supplies and equipment using the Leonardo module. Again, the crew deployed to the ISS have their names in both Roman and Cyrillic letters at the bottom on the mission patch.

STS-108 (ENDEAVOUR)

LAUNCH DATE December 5, 2001
FLIGHT TIME 11 days, 19 hours, 36 minutes
CREW Dominic Gorie, Mark Kelly, Linda Godwin, Daniel Tani

Endeavour docked to the station and resupplied it from the Raffaello module, as well as swapping ISS crews. The construction of the ISS is emphasized in this busy STS-108 mission patch design.

STS-109 (COLUMBIA)

LAUNCH DATE March 1, 2002
FLIGHT TIME 10 days, 22 hours, 11 minutes
CREW Scott Altman, Duane Carey, John Grunsfeld, Nancy Currie, Richard Linnehan, James Newman, Michael Massimino

The STS-109 crew successfully serviced the Hubble Space Telescope. The upgrades and servicing by the crew placed a new power unit, a new camera, and new solar arrays in the Hubble. Hubble servicing takes center stage in this elongated oval patch.

STS-110 (ATLANTIS)

LAUNCH DATE April 8, 2002

FLIGHT TIME 10 days, 19 hours, 43 minutes

CREW Michael Bloomfield, Stephen Frick, Rex Walheim, Ellen Ochoa, Lee Morin, Jerry Ross, Steven Smith

During STS-110, the crew undertook construction of the ISS, delivering and installing the S0 truss required for future module attachment as well as cooling and power systems. This unglamorous work required four space walks. More abstract in design than most other mission patches, this one also focused on the ISS construction element of the mission.

STS-111 (ENDEAVOUR)

LAUNCH DATE June 5, 2002

FLIGHT TIME 13 days, 20 hours, 36 minutes

CREW Kenneth Cockrell, Paul Lockhart, Franklin Chang-Diaz, Philippe Perrin

This crew delivered the multipurpose logistics module Leonardo to the ISS, containing more than 5,600 pounds of supplies and equipment. It also affixed the mobile remote service base system to the ISS, enabling the Canadarm 2 to move up and down the station. A uniquely satisfying and clean design, the mission patch features the ISS in orbit below a shuttle rising to the top of the astronaut emblem.

STS-112 (ATLANTIS)

LAUNCH DATE October 7, 2002
FLIGHT TIME 10 days, 19 hours, 59 minutes
CREW Jeffrey Ashby, Pamela Melroy, David Wolf, Sandra Magnus, Piers Sellers, Fyodor Yurchikhin

STS-112 carried the S1 integrated truss segment and the crew and equipment translation aid, providing mobile work platforms for future spacewalking astronauts. Three space walks by David Wolf and Piers Sellers accomplished this installation in 19 hours and 41 minutes. This octagonal mission patch takes a different graphic approach with no depictions of the space shuttle. Instead it emphasizes the ISS construction effort.

STS-113 (ENDEAVOUR)

LAUNCH DATE November 23, 2002
FLIGHT TIME 13 days, 18 hours, 49 minutes
CREW James Wetherbee, Paul Lockhart, Michael Lopez-Alegria, John Herrington

Swapping the Expedition 5 crew with the Expedition 6 astronauts, this mission also installed trusses on the ISS. Three space walks took place to outfit and activate the truss. Garish in the use of red and yellow colors, and in the use of Roman numerals for the mission number, the swooshing astronaut emblem dominated the center of the patch. Once again, the ISS crew are named in both languages.

STS-107 (COLUMBIA)

LAUNCH DATE January 16, 2003
FLIGHT TIME 15 days, 22 hours, 39 minutes
CREW Rick Husband, William McCool, Michael Anderson, David Brown, Kalpana Chawla, Laurel Clark, Ilan Ramon

STS-107 ended tragically on February 1, 2003, when the *Columbia* broke up during reentry 16 minutes before a scheduled landing at the Kennedy Space Center. NASA's Mission Control Center abruptly lost both telemetry and communication with the spacecraft as it proceeded through the atmosphere on its way to a normal landing. Investigators later learned that a hole blasted into the leading edge of the right shuttle wing had allowed superheated air to damage the orbiter's superstructure. It broke up at about 216,000 feet, and the entire crew was lost during this tragic event. In the aftermath, NASA stood down all shuttle flights for 29 months. It modified the shuttles for a return to flight, and revamped safety procedures and operational requirements.

The mission patch is one of the most famous in the history of the shuttle program. It is the only one shaped like an orbiter, but more importantly for the ISS era, it emphasizes scientific research rather than ISS construction. There are subtle ways in which this was done. The symbol for microgravity, μg, is emblazoned on the patch at the bottom to highlight the microgravity research aspects of this flight. The crew also included Ilan Ramon, the first Israeli astronaut, and one of the stars depicted on the patch has six points like a Star of David, symbolizing the Israel Space Agency's contributions to the mission.

145

STS-114 (DISCOVERY)

LAUNCH DATE July 26, 2005

FLIGHT TIME 13 days, 21 hours, 33 minutes

CREW Eileen Collins, James Kelly, Soichi Noguchi, Stephen Robinson, Andrew Thomas, Wendy Lawrence, Charles Camarda

After nearly 30 months without flights on the space shuttle, the STS-114 mission marked the return to flight for the shuttle fleet. During the interim, NASA had worked to determine and correct the technical problems that had led to the loss of *Columbia* and its crew. The ISS had been sustained in orbit by Russian resupply and crew rotations during this hiatus. STS-114 signaled the forthcoming ending of the shuttle program; U.S. president George W. Bush had announced in January 2004 that the venerable shuttle would be retired after the completion of the ISS.

A satisfying nonlinear design, this mission patch shows a tiny shuttle at the top with seven stars for the lost crew of *Columbia*. This patch, in recognition of the partnership with the Japanese Aerospace Exploration Agency (JAXA), included a red sun in honor of Japanese contributions. It also features the ISS, which is represented on the blue orbit around Earth.

STS-121 (DISCOVERY)

LAUNCH DATE July 4, 2006

FLIGHT TIME 12 days, 18 hours, 38 minutes

CREW Steven Lindsey, Mark Kelly, Michael Fossum, Lisa Nowak, Stephanie Wilson, Piers Sellers, Thomas Reiter

It took nearly a year after the return to flight for NASA to resume full-scale construction of the ISS using the space shuttle. STS-121 represented a bit of a return to normalcy for the space agency's activities. This crew docked at the station and transferred 7,400 pounds of supplies and equipment from the multipurpose logistics module Leonardo to the ISS, even as they undertook three space walks totaling more than 21 hours. During that time, crew members installed several components on the ISS trusses, repaired items that had malfunctioned, and made ready for additional construction efforts in future shuttle flights. Finally, because of concerns about possible damage to orbiters during flight, NASA inaugurated a set of imaging procedures to ensure that any anomalies would not be overlooked. Photography of the shuttle from multiple angles was thereafter a part of every flight.

The STS-121 mission patch was simple and elegant, depicting the shuttle docked to the ISS at the center. The astronaut emblem creates a striking feature, catching the eye and drawing it upward.

STS-115 (ATLANTIS)

LAUNCH DATE September 9, 2006
FLIGHT TIME 11 days, 19 hours, 6 minutes
CREW Brent Jett, Christopher Ferguson, Joseph Tanner, Daniel Burbank, Heidemarie Stefanyshyn-Piper, Steven MacLean

Atlantis delivered truss section P3/P4, along with other supplies to the station. A busy mission patch, the solar array deployed in the foreground emphasizes the chief construction effort of this mission.

STS-116 (DISCOVERY)

LAUNCH DATE December 9, 2006
FLIGHT TIME 12 days, 20 hours, 45 minutes
CREW Mark Polansky, William Oefelein, Nicholas Patrick, Robert Curbeam, Joan Higginbotham, Christer Fuglesang, Sunita Williams, Thomas Reiter

This mission included the first station crew rotation by a shuttle in four years. The crew also installed the P5 integrated truss segment, nearly doubling the electrical power available for ISS operations. The oval mission patch includes the shuttle swooping upward with American and Swedish flags trailing behind; Christer Fuglesang was an ESA astronaut from Sweden.

STS-117 (ATLANTIS)

LAUNCH DATE June 8, 2007

FLIGHT TIME 13 days, 20 hours, 13 minutes

CREW Frederick Sturckow, Lee Archambault, Patrick Forrester, Steven Swanson, John "Danny" Olivas, James Reilly, Clayton Anderson

STS-117 continued the unexciting, but necessary, construction of the S3/S4 truss segment for the ISS and the solar arrays and other components necessary for operation. The crew patch was a baroque, almost gothic design complete with stars, stripes, banners, and drapings that showed the ISS in the center.

STS-118 (ENDEAVOUR)

LAUNCH DATE August 8, 2007

FLIGHT TIME 12 days, 17 hours, 56 minutes

CREW Scott Kelly, Charles Hobaugh, Tracy Caldwell Dyson, Richard Mastracchio, Dafydd Williams, Barbara Morgan, Alvin Drew

This was the first flight of *Endeavour* in three years. It provided the first use of the Station-to-Shuttle Power Transfer System. With a color palette using blue, black, yellow, and red, this mission patch shows a shuttle swooping through the foreground trailing the Stars and Stripes with the ISS in silhouette in the background.

STS-120 (ATLANTIS)

LAUNCH DATE October 23, 2007

FLIGHT TIME 15 days, 2 hours, 24 minutes

CREW Pamela Melroy, George Zamka, Douglas Wheelock, Stephanie Wilson, Scott Parazynski, Paolo Nespoli, Daniel Tani, Clayton Anderson

During the mission, the STS-120 crew continued construction of the ISS with the installation of the Harmony Node 2 module and relocating the P6 truss. The mission patch features a subdued shuttle silhouette and the star in the distance that is the ISS. It also has to its center right the Moon and Mars, which the crew emphasized was the future of human space exploration.

STS-122 (ATLANTIS)

LAUNCH DATE February 7, 2008

FLIGHT TIME 12 days, 18 hours, 22 minutes

CREW Stephen Frick, Alan Poindexter, Leland Melvin, Rex Walheim, Hans Schlegel, Stanley Love, Léopold Eyharts, Daniel Tani

STS-122 delivered and installed the European Space Agency's Columbus laboratory, whose experiments were to be coordinated by the Columbus Control Centre in Oberpfaffenhofen, Germany. This mission patch emphasizes the laboratory, shown here with a representation of Christopher Columbus's 1492 ship in the foreground. The crew wanted to depict that 500 years after Columbus sailed to America, the STS-122 crew took the laboratory module to the ISS to usher in a new era of scientific discovery.

STS-123 (ENDEAVOUR)

LAUNCH DATE March 11, 2008

FLIGHT TIME 15 days, 18 hours, 12 minutes

CREW Dominic Gorie, Gregory Johnson, Robert Behnken, Michael Foreman, Richard Linnehan, Takao Doi, Garrett Reisman, Léopold Eyharts

Endeavour's record-breaking 12-day stay at the ISS included five space walks. The shuttle crew, working with the Expedition 16 members, attached the first section of the Japanese experiment module to the station. With the astronaut symbol as a central feature on the patch, the space shuttle is seen constructing the ISS. The Dextre robotic system is also shown on the left. The space shuttle is seen in orbit with the crew names trailing behind.

STS-124 (DISCOVERY)

LAUNCH DATE May 31, 2008

FLIGHT TIME 13 days, 18 hours, 13 minutes

CREW Mark Kelly, Kenneth Ham, Karen Nyberg, Ronald Garan, Michael Fossum, Akihiko Hoshide, Garrett Reisman, Greg Chamitoff

STS-124 delivered the pressurized module and robotic arm of the Japanese experiment module (JEM), known as "Kibo" (meaning "hope"), to the ISS. This joint U.S.–Japan mission to the ISS takes center stage in this dramatically colored mission patch. The JEM appears center stage in this emblem. The crew indicated that the Sun's rays in the background signified the promise of scientific knowledge gained through research using Kibo.

STS-126 (ENDEAVOUR)

LAUNCH DATE November 14, 2008

FLIGHT TIME 15 days, 20 hours, 30 minutes

CREW Christopher Ferguson, Eric Boe, Stephen Bowen, Robert Kimbrough, Sandra Magnus, Donald Pettit, Heidemarie Stefanyshyn-Piper, Greg Chamitoff

Endeavour delivered equipment to the ISS that enabled larger crews to reside aboard the complex, especially the Leonardo multipurpose logistics module. With the astronaut emblem at top center, and the shuttle swooshing to and from the ISS, the STS-126 patch uses subdued colors to identify the mission and depict the logistics module.

STS-119 (DISCOVERY)

LAUNCH DATE March 15, 2009

FLIGHT TIME 12 days, 19 hours, 30 minutes

CREW Lee Archambault, Dominic "Tony" Antonelli, Joseph Acaba, John Phillips, Steve Swanson, Richard Arnold, Koichi Wakata, Sandra Magnus

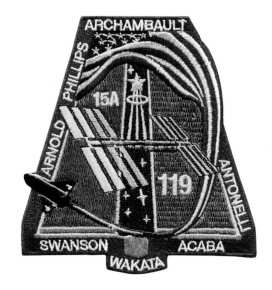

STS-119 assembled the S6 truss, a final set of U.S. solar arrays, and a distillation assembly to get the station's water recycling system into operation. The crew sought to depict the solar array with their mission patch, draping it in the Stars and Stripes and showing the ISS in the center. Once again, the astronaut emblem appears in this patch, this time at the center.

STS-125 (ATLANTIS)

LAUNCH DATE May 11, 2009
FLIGHT TIME 12 days, 21 hours, 37 minutes
CREW Scott Altman, Gregory Johnson, John Grunsfeld, Michael Massimino, Andrew Feustel, Michael Good, Megan McArthur

This fourth servicing of the Hubble Space Telescope required five space walks. The crew installed two new instruments and repaired two others, bringing them back to life; replaced gyroscopes and batteries; and added new thermal insulation panels to protect the orbiting observatory. The STS-125 crew played up the servicing of the telescope and the portrait of the universe available because of its continued operation. Uniquely, the patch suggests that our knowledge about galaxies, stars, and planets is widened through the telescope.

STS-127 (ENDEAVOUR)

LAUNCH DATE July 15, 2009
FLIGHT TIME 15 days, 16 hours, 45 minutes
CREW Mark Polansky, Douglas Hurley, Christopher Cassidy, Julie Payette, Thomas Marshburn, David Wolf, Timothy Kopra

With a uniquely dark palette of colors, the oval mission patch depicts the ISS construction in an abstract manner. Its crew emphasized, "Bathed in sunlight, the blue Earth is represented without boundaries to remind us that we all share this world. In the center, the golden flight path of the space shuttle turns into the three distinctive rays of the astronaut symbol culminating in the star-like emblem characteristic of the Japanese Aerospace Agency, yet soaring further into space as it paves the way for future voyages and discoveries for all humankind."

STS-128 (DISCOVERY)

LAUNCH DATE August 28, 2009
FLIGHT TIME 13 days, 20 hours, 54 minutes
CREW Frederick Sturckow, Kevin Ford, Patrick Forrester, José Hernández, John "Danny" Olivas, Christer Fuglesang, Nicole Stott

This was the 30th mission of a space shuttle dedicated to the assembly and maintenance of the ISS. Uniquely shaped, the mission insignia is dominated by the astronaut emblem in the center, as well as the shuttle orbiter and the ISS. Earth and the ISS wrap around the astronaut emblem, foreshadowing the continuous human presence in space. The names of the crew border the patch in an unfurled banner, with the U.S. and Swedish flags on either side.

STS-129 (ATLANTIS)

LAUNCH DATE November 16, 2009
FLIGHT TIME 10 days, 19 hours, 16 minutes
CREW Charles Hobaugh, Barry Wilmore, Leland Melvin, Randolph Bresnik, Michael Foreman, Robert Satcher, Nicole Stott

This crew delivered critical equipment and supplies to the ISS, as well as undertaking three space walks to complete elements of the station's construction. In one of the most extravagant mission patch designs, the STS-129 crew used the astronaut emblem as the centerpiece along with the shuttle orbiter, the ISS, and both the Moon and Mars. It also depicted 13 stars on the patch, to symbolize the crew's children.

STS-130 (ENDEAVOUR)

LAUNCH DATE February 8, 2010
FLIGHT TIME 13 days, 18 hours, 6 minutes
CREW George Zamka, Terry Virts, Kathryn Hire, Stephen Robinson, Nicholas Patrick, Robert Behnken

STS-130 carried the Tranquility node and the Cupola, a windowed robotics viewing station from which astronauts have the opportunity not only to monitor a variety of the station operations but also to study Earth. The mission patch reflects this emphasis, using the hexagonal shape of the Cupola as its central design feature. The orbiter swooshes through the frame with Earth in the background, an image reminiscent of the famous view of Earth taken by *Lunar Orbiter 1* on August 23, 1966.

STS-131 (DISCOVERY)

LAUNCH DATE April 5, 2010
FLIGHT TIME 15 days, 2 hours, 47 minutes
CREW Alan Poindexter, James Dutton, Richard Mastracchio, Dorothy Metcalf-Lindenburger, Stephanie Wilson, Naoko Yamazaki, Clayton Anderson

In one of the last ISS construction flights, this crew patch highlights the space shuttle during a new post-*Columbia* maneuver that allowed the imaging of the shuttle exterior prior to reentry. Depicted in the space shuttle's cargo bay is the Leonardo module, which was taken to the ISS with science racks, equipment, and supplies. The patch also uses as a key feature the astronaut emblem centered at a 45-degree angle.

STS-132 (ATLANTIS)

LAUNCH DATE May 14, 2010
FLIGHT TIME 11 days, 18 hours, 29 minutes
CREW Kenneth Ham, Dominic "Tony" Antonelli, Garrett Reisman, Michael Good, Stephen Bowen, Piers Sellers

STS-132 was initially scheduled to be the final flight of *Atlantis*, but NASA leaders extended the program to include STS-135 in 2011. In this mission to the ISS, STS-132 delivered an Integrated Cargo Carrier and a Russian-built Mini Research Module-1 (MRM-1), named "Rassvet," the Russian word for "dawn." In the mission patch, the shuttle is shown pointing away from the viewer and toward a setting sun, symbolizing the near-end of the space shuttle program.

STS-133 (DISCOVERY)

LAUNCH DATE February 24, 2011
FLIGHT TIME 12 days, 19 hours, 5 minutes
CREW Steven Lindsey, Eric Boe, Nicole Stott, Alvin Drew, Michael Barratt, Stephen Bowen

Based upon sketches from the late legendary artist Robert McCall, in this patch's foreground is *Discovery* ascending to orbit on a plume of flame in front of a top-lit Earth crescent and star field beyond. Uniquely, this launch shows only the orbiter, without boosters or an external tank, as it would be at mission's end. This is to signify *Discovery*'s completion of its operational life and the beginning of its new role as a symbol of NASA and the nation's proud legacy in human spaceflight.

STS-134 (ENDEAVOUR)

LAUNCH DATE May 16, 2011
FLIGHT TIME 15 days, 17 hours, 39 minutes
CREW Mark Kelly, Gregory Johnson, Michael Fincke, Roberto Vittori, Andrew Feustel, Gregory Chamitoff

Endeavour delivered one of the last parts of the ISS, the Alpha Magnetic Spectrometer, during the program's 36th shuttle mission. The concentric ovals of this mission patch offered an elegant portrayal of the crew as individuals working together to complete their ISS construction flight. They based their design on the symbol of the atom, with their names serving as electrons around a nucleus.

STS-135 (ATLANTIS)

LAUNCH DATE July 8, 2011
FLIGHT TIME 12 days, 18 hours, 29 minutes
CREW Christopher Ferguson, Douglas Hurley, Sandra Magnus, Rex Walheim

During this last space shuttle mission, there was only a four-person crew, which was the smallest of any shuttle flight since STS-6 in April 1983. It delivered the Raffaello multipurpose logistics module containing supplies and equipment for the ISS. The STS-135 mission patch encapsulates design elements from many other missions along with a version of the NASA logo in the center. It depicts the space shuttle *Atlantis* during launch, flying through the Greek omega symbol, suggesting the end of the 30-year program.

CHAPTER 9

ISS EXPEDITIONS 2000–2010 ⟨⟨⟩⟩

The International Space Station (ISS) flies over Earth during the approach of STS-105 on August 12, 2001. This mission to construct the ISS rotated the Expedition 2 crew for Expedition 3 and brought critical supplies and equipment. At this stage of construction the ISS had only three major modules, trusses, and a rudimentary solar array system for the generation of electrical power.

ISS EXPEDITIONS 2000–2010

In 1984, NASA began the international effort to build a space station. Together with a consortium of global partners, they set about a series of negotiations to decide its scope and design, which organizations would be responsible for which parts, and how its operation would be organized. Eventually 15 nations participated in the ISS program, including the U.S., Canada, Japan, Russia, Belgium, Denmark, France, Germany, Italy, the Netherlands, Norway, Spain, Sweden, Switzerland, and the United Kingdom.

Russia was not brought into the space station partnership until the end of the Cold War. In the early 1990s, NASA opened negotiations with Roscosmos in which the human exploration elements of both programs developed a symbiotic relationship. The Americans had the space shuttle, and the Russians had the *Mir* space station and a serviceable Soyuz launch vehicle. Both had extensive human spaceflight experience that strengthened the capabilities of both nations.

The ISS may well be recognized in the future as the start of a merger of the world's spacefaring nations to undertake great explorations of the solar system. Building the station was a remarkably difficult and complex technological endeavor, and its partner nations have been able to maintain their relationships regardless of the vicissitudes of international politics. From the joining of the first two station components, the Zarya module from Roscosmos, and NASA's Unity module in 1998, through the completion of the station in 2011, the success of this program has been obvious. With the service life of the ISS now anticipated to last until 2028, this may well be the longest, most complex, and uniquely rewarding peacetime international program in history.

1]
The *Soyuz TMA-5* vehicle was rolled to its launch pad at the Baikonur Cosmodrome in Kazakhstan on October 12, 2004, in preparation for its launch on October 14, to send Expedition 10's Leroy Chiao, Salizhan Sharipov, and Yuri Shargin to the International Space Station.

2]
The ISS was an unparalleled international scientific and technological cooperative venture. It began on December 4, 1998, with the assembly of the first two modules. This schematic shows the modules of the ISS at completion.

The formal operational life of the International Space Station began on October 31, 2000, when the first crew arrived. Expedition 1, consisting of American astronaut Bill Shepherd and Russian cosmonauts Yuri Gidzenko and Sergei Krikalev, flew on a Russian Soyuz from Baikonur Cosmodrome, Kazakhstan. After arriving on the ISS, they began to put operations in order. They welcomed assembly crews arriving by space shuttle, and a routine developed consisting of expedition crew rotations and additional components arriving that had to be assembled and attached to the existing modules. Progress was mundane but steady until February 1, 2003. On that date the space shuttle *Columbia* broke up on reentry from space and its crew of seven astronauts died. NASA had no choice but to stand down shuttle operations until the problems could be identified and corrected. The grounded space shuttle fleet put the assembly of the ISS on hiatus for some 30 months.

Regardless, crews stayed aboard the ISS to keep it operational. While there had been three astronauts/cosmonauts aboard prior to the *Columbia* accident, the ISS partners reduced the crew to two during this period to conserve supplies, which continued to be brought up by Russian Soyuz spacecraft. Only after the return of the space shuttle to flight on July 26, 2005, did crew numbers aboard the ISS rise once again. At first it was three, but soon six astronauts and cosmonauts worked together on the orbital platform. The shuttle also began ferrying new modules and other components to the ISS beginning in 2005, starting the run that would see the functional completion of the station by 2011.

Over the course of the ISS's history until the beginning of 2020, 61 crews—also called expeditions—have served at the station. Each crew designed their own mission patch that became an individualized statement of their mission.

3]
Astronaut James Voss, Expedition 2, tends to experiments in the Destiny laboratory of the ISS on July 17, 2002. Behind him, astronaut Scott Horowitz, mission commander of STS-105, floats through the hatchway leading from the Unity node.

4]
On February 10, 2001, during STS-98, the crew of the space shuttle *Atlantis* deliver the Destiny laboratory module to the ISS. Here, the space shuttle's Canadarm remote manipulator system is removed from the payload bay and prepared for mating to the ISS.

5]
STS-127 astronaut Thomas Marshburn undertakes a space walk in July 2009 to affix the Kibo Japanese Experiment Module to the ISS.

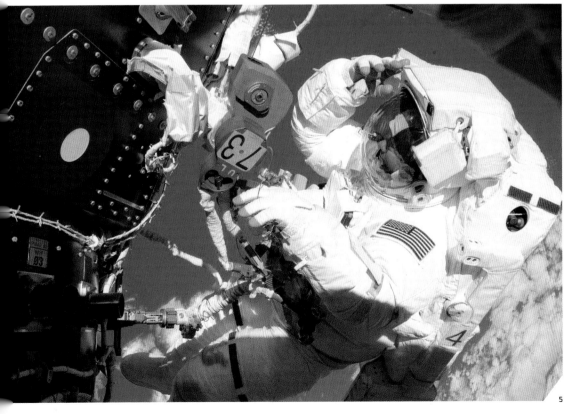

EXPEDITION 1

DATE October 31, 2000–March 21, 2001
DOCKING November 2, 2000
UNDOCKING March 19, 2001
VEHICLES *Soyuz TM-31* (launch), *Discovery* (STS-102) (return)
CREW William Shepherd, Yuri Gidzenko, Sergei Krikalev

Expedition 1 marked the first crew to inhabit the ISS. The artwork for the first ISS crew patch is a simplified graphic of the station complex when fully completed.

EXPEDITION 2

DATE March 8–August 22, 2001
DOCKING March 10, 2001
UNDOCKING August 20, 2001
VEHICLES *Discovery* (STS-102) (launch), *Discovery* (STS-105) (return)
CREW Yury Usachev, James Voss, Susan Helms

Expedition 2 saw further construction of the ISS during four space shuttle missions and one Soyuz mission to the outpost. The patch depicts the ISS as it appeared at the time the second crew went aboard. The stars in the background, according to the crew, "represent the thousands of space workers throughout the ISS partnership."

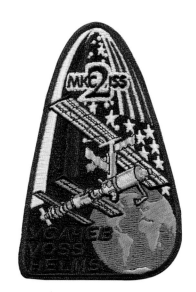

EXPEDITION 3

DATE August 10–December 17, 2001
DOCKING August 12, 2001
UNDOCKING December 15, 2001
VEHICLES *Discovery* (STS-105) (launch), *Endeavour* (STS-108) (return)
CREW Frank Culbertson, Vladimir Dezhurov, Mikhail Tyurin

In four space walks, the Expedition 3 crew continued with the on-orbit construction and maintenance of the ISS, including repairing an obstruction that prevented a Progress supply ship from docking with the station. Expedition 3 was in space during the September 11, 2001, terrorist attacks, and crew members took photos of the devastation at the World Trade Center site.

The Expedition 3 crew members said about this insignia: "The book of space history turns from the chapter written onboard the Russian *Mir* station and the U.S. space shuttle to the next new chapter, one that will be written on the blank pages of the future by space explorers working for the benefit of the entire world. The space walker signifies the human element of this endeavor. The star representing the members of the third expedition, and the entire multi-national space station building team, streaks into the dawning era of cooperative space exploration, represented by the image of the ISS as it nears completion."

EXPEDITION 4

DATE December 5, 2001–June 19, 2002
DOCKING December 7, 2001
UNDOCKING June 15, 2002
VEHICLES *Endeavour* (STS-108) (launch), *Endeavour* (STS-111) (return)
CREW Yury Onufriyenko, Daniel Bursch, Carl Walz

This crew saw Carl Walz and Daniel Bursch break the U.S. spaceflight endurance record of 188 days by spending 231 days in space. The diamond shape of the crew patch signified the "diamond in the rough" nature of the ISS at this stage, pointing to the challenges the crew faced. The patch also shows the flags of the U.S. and Russia, highlighting the international nature of the mission.

EXPEDITION 5

DATE June 5–December 7, 2002
DOCKING June 7, 2002
UNDOCKING December 2, 2002
VEHICLES *Endeavour* (STS-111) (launch), *Endeavour* (STS-113) (return)
CREW Valery Korzun, Peggy Whitson, Sergei Treschev

Expedition 5 crew members inhabited the ISS during a time of relatively few additions to the station and conducted only two space walks. But during this mission the ISS was visited by three space shuttle crews. Crew members emphasized the international elements of the ISS in this patch design, complete with bunting made from U.S. and Russian flags. The main feature of the patch depicts the shuttle docked to the ISS.

EXPEDITION 6

DATE November 23, 2002–May 4, 2003
DOCKING November 25, 2002
UNDOCKING May 3, 2003
VEHICLES *Endeavour* (STS-113) (launch), *Soyuz TMA-2* (return)
CREW Kenneth Bowersox, Donald Pettit, Nikolai Budarin

During Expedition 6's time in space, *Columbia* broke apart during reentry, grounding all shuttle missions and forcing the ISS to rely on Russian vehicles. Due to this accident, Expedition 6 returned aboard *Soyuz TMA-2*. This crew patch depicts the ISS orbiting our planet to emphasize the work being conducted onboard to improve life on Earth.

EXPEDITION 7

DATE April 25–October 27, 2003
DOCKING April 28, 2003
UNDOCKING October 27, 2003
VEHICLE *Soyuz TMA-2* (launch and return)
CREW Yuri Malenchenko, Edward Lu

Expedition 7 marked the first two-person crew to inhabit the ISS. This change resulted from the grounding of the shuttle fleet, as crew members Yuri Malenchenko and Edward Lu spent almost 200 days in space. This patch seeks to honor the American and Russian space programs, and how they came together in the ISS effort, suggested by the two elliptical orbits.

EXPEDITION 8

DATE October 18, 2003–April 29, 2004
DOCKING October 20, 2003
UNDOCKING April 29, 2004
VEHICLE *Soyuz TMA-3* (launch and return)
CREW Michael Foale, Aleksandr Kaleri

Expedition 8 launched with three crew members, though only two of them stayed on the station. This mission patch plays up the U.S.-Russian aspects of the flight with flags of each nation making something of a figure eight around Earth and the ISS above.

EXPEDITION 9

DATE April 18–October 23, 2004
DOCKING April 21, 2004
UNDOCKING October 23, 2004
VEHICLE *Soyuz TMA-4* (launch and return)
CREW Gennady Padalka, Michael Fincke

While three launched on the *Soyuz TMA-4* to the ISS, only two stayed as the Expedition 9 crew. With the space shuttle fleet still out of service, the crew of Expedition 9 continued to focus on the ISS operations and scientific research. One of the busiest mission patches of the ISS, this one shows the Soyuz launcher and an X, forming the Roman numeral IX, the ISS, and a star-spangled eagle in flight.

EXPEDITION 10

DATE October 13, 2004–April 24, 2005
DOCKING October 16, 2004
UNDOCKING April 24, 2005
VEHICLE *Soyuz TMA-5* (launch and return)
CREW Leroy Chiao, Salizhan Sharipov

During the 2004 U.S. presidential election, Leroy Chiao became the first astronaut to vote in space, making his choice through an electronic ballot sent to the station. The Expedition 10 patch has a large Roman numeral X, formed by the U.S. and Russian flags. A graphic of the current configuration of the ISS is shown on the left.

EXPEDITION 11

DATE April 14–October 10, 2005
DOCKING April 16, 2005
UNDOCKING October 10, 2005
VEHICLE *Soyuz TMA-6* (launch and return)
CREW Sergei Krikalev, John Phillips

Expedition 11 was commanded by cosmonaut Sergei Krikalev, who broke the previous 749-day record for most time spent in space by a human. Expedition 11's mission patch is explained by the crew: "The beauty of our home planet and the vivid contrasts of the space environment are shown by the blue and green Earth with the ISS orbiting overhead, and by the bright stars, dark sky, and dazzling Sun."

EXPEDITION 12

DATE September 30, 2005–April 8, 2006
DOCKING October 3, 2005
UNDOCKING April 8, 2006
VEHICLE Soyuz TMA-7 (launch and return)
CREW William McArthur, Valery Tokarev

American space tourist Gregory Olsen joined the Expedition 12 crew as they launched to the ISS, and then returned with the Expedition 11 crew after a brief stay. The Expedition 12 crew patch featured the ISS prominently, emphasizing the hope that it would continue to grow in its capability as a world-class laboratory and test bed for exploration. The crew added that the vision of exploration is depicted by the Moon and Mars.

EXPEDITION 13

DATE March 29–September 28, 2006
DOCKING March 31, 2006
UNDOCKING September 28, 2006
VEHICLE Soyuz TMA-8 (launch and return)
CREW Pavel Vinogradov, Jeffrey Williams, Thomas Reiter

Though launched with only two crew members, Expedition 13 received a third astronaut from STS-121, the first space shuttle mission to the ISS following the *Columbia* accident in 2003, returning the ISS crew complement to three. The ISS Expedition 13 crew commented on their mission patch: "The dynamic trajectory of the space station against the background of the Earth, Mars, and the Moon symbolizes the vision for human space exploration beyond Earth orbit and the critical role that the ISS plays in the fulfillment of that vision."

EXPEDITION 14

DATE September 17, 2006–April 21, 2007
DOCKING September 20, 2006
UNDOCKING April 21, 2007
VEHICLE *Soyuz TMA-9* (launch and return)
CREW Michael Lopez-Alegria, Mikhail Tyurin, Thomas Reiter, Sunita Williams

A second space tourist to the ISS, American businesswoman Anousheh Ansari, launched with Expedition 14 and returned to Earth with the former crew. Additionally, Thomas Reiter remained on the ISS. In an elegant design, the Expedition 14 patch featured the Roman numeral XIV suspended above the Earth as a reminder of this mission and its trajectory leading toward exploration of the Moon, Mars, and beyond. The crew wanted to commemorate those lost during the *Apollo 1*, *Soyuz 1*, *Soyuz 11*, *Challenger*, and *Columbia* missions with the five stars on the patch.

EXPEDITION 15

DATE April 7, 2007–October 21, 2007
DOCKING April 9, 2007
UNDOCKING October 21, 2007
VEHICLE *Soyuz TMA-10* (launch and return)
CREW Fyodor Yurchikhin, Oleg Kotov, Clayton Anderson

During Expedition 15, the ISS Integrated Truss Structure was expanded twice: STS-117 brought the S3/S4 truss, and STS-118 brought the S5 truss. The relationship of flight controllers and the crew in space was reflected in the Expedition 15 mission patch. Against a backdrop familiar on the big board of Mission Control, the expedition is represented in Roman numerals over a Mercator projection of Earth. The ISS is shown in a fully operational configuration, although it looked nothing like this at the time of Expedition 15.

EXPEDITION 16

DATE October 10, 2007–April 19, 2008
DOCKING October 12, 2007
UNDOCKING April 19, 2008
VEHICLE *Soyuz TMA-11* (launch and return)
CREW Peggy Whitson, Yuri Malenchenko, Clayton Anderson, Garrett Reisman

Expedition 16 saw an important first in the ISS program. This was the first time a crew served twice on the ISS, as Peggy Whitson and Yuri Malenchenko had served together on Expedition 7. Additionally, Malenchenko had previously been an ISS commander, but now returned as a flight engineer. Whitson was also the first female commander of an ISS expedition, and when STS-120 arrived commanded by female astronaut Pamela Melroy, it marked the first time that two women mission commanders were present at the same time.

EXPEDITION 17

DATE April 8–October 23, 2008
DOCKING April 10, 2008
UNDOCKING October 24, 2008
VEHICLE *Soyuz TMA-12* (launch and return)
CREW Sergey Volkov, Oleg Kononenko, Garrett Reisman, Gregory Chamitoff

The Expedition 17 crew spent the majority of their time working to ensure new modules and systems functioned properly as the ISS took shape. Their mission patch, like so many before, was rich in symbolism. It focused on the potential of using the ISS, shown at the bottom of the patch, as a place to learn what was necessary to survive in deep space and a place from which to expand outward in the quest to reach other bodies in the solar system and beyond.

EXPEDITION 18

DATE October 12, 2008–April 8, 2009
DOCKING October 14, 2008
UNDOCKING April 7, 2009
VEHICLE *Soyuz TMA-13* (launch and return)
CREW Michael Fincke, Yuri Lonchakov, Sandra Magnus, Koichi Wakata, Greg Chamitoff

On March 12, 2009, a piece of debris from a Delta II launch vehicle upper stage passed close to the ISS, forcing the crew to board a Soyuz capsule docked to the ISS in the event the station was hit. Fortunately, the debris did not hit the station and the crew resumed operations. The Expedition 18 crew explained that the Roman numeral XVIII on their crew patch designated not only their mission number, but sought to use the "X" to evoke exploration, the "V" to stand for victory (and also the five space agencies in the ISS effort), and the "III" representing the number in the crew.

EXPEDITION 19

DATE March 26–May 29, 2009
DOCKING March 28, 2009
UNDOCKING May 29, 2009
VEHICLE *Soyuz TMA-14* (launch and return)
CREW Gennady Padalka, Michael Barratt, Koichi Wakata

Expedition 19 was the final expedition with three crew members, before the crew size increased to six with Expedition 20. This expedition saw a disagreement over the shared use of nationally supplied facilities, especially exercise equipment and bathrooms. Russian and American crew members were directed to use only their own supplies. This disagreement passed quickly, but it led to a general lowering of morale on the ISS for several months. The crew patch of Expedition 19 featured the Earth prominently.

EXPEDITION 20

DATE May 29–October 10, 2009
DOCKING May 29, 2009
UNDOCKING October 10, 2009
VEHICLES *Soyuz TMA-14, Soyuz TMA-15* (launch and return)
CREW Gennady Padalka, Michael Barratt, Koichi Wakata, Frank De Winne, Roman Romanenko, Robert Thirsk

Construction continued during this expedition, but the most publicized activity was a special experiment by Koichi Wakata wherein he did not change his underpants for one month, testing a specially designed garment that did not need washing or changing. He reported no undue odor or effects on his body. The Expedition 20 mission patch symbolized the advent of the six-person crew onboard the ISS with six gold stars. The astronaut symbol also extends from the base to the top.

As the crew noted: "The space station in the foreground represents where we are now and the important role it is playing towards meeting our exploration goals. The knowledge and expertise developed from these advancements will enable us to once again leave low Earth orbit for the new challenges of establishing a permanent presence on the Moon and then on to Mars."

EXPEDITION 21

DATE September 30–December 1, 2009
DOCKING May 29, 2009
UNDOCKING December 1, 2009
VEHICLES Soyuz TMA-15, Soyuz TMA-16, STS-128 (launch and return), STS-129 (launch)
CREW Frank De Winne, Roman Romanenko, Robert Thirsk, Nicole Stott, Jeffrey Williams, Maksim Surayev

The crew used the new Russian Mini-Research Module 2 for experimental work on the ISS. During the handover between Expedition 20 and 21, three Soyuz vehicles were docked to the ISS at the same time, a first for the station. According to the crew, "the central element of this patch was inspired by a fractal of six, symbolizing the six-person crew. The patch also shows children as our future and a reason to explore."

EXPEDITION 22

DATE September 30, 2009–March 18, 2010
DOCKING October 2, 2009
UNDOCKING March 18, 2010
VEHICLES Soyuz TMA-16, Soyuz TMA-17 (launch and return)
CREW Jeffrey Williams, Maksim Surayev, Oleg Kotov, Soichi Noguchi, Timothy Creamer

This expedition crew continued experiments and prepared the Poisk module for future dockings, using Russian-built Orlan space suits for the first time. After three members of the Expedition 21 crew departed on *Soyuz TMA-16*, a period of three weeks followed when only two crew members remained, the first time that had happened since Expedition 13. Expedition 22's mission patch emphasized the final assembly of the ISS and its use for research. The six stars illustrated the increased capability of the crew complement.

EXPEDITION 23

DATE March 18–June 1, 2010
DOCKING December 22, 2009
UNDOCKING June 1, 2010
VEHICLES *Soyuz TMA-17*, *Soyuz TMA-18* (launch and return)
CREW Oleg Kotov, Soichi Noguchi, Timothy Creamer, Aleksandr Skvortsov, Mikhail Korniyenko, Tracy Caldwell Dyson

This was the first crew to include three Russians on the ISS at the same time. This crew continued outfitting the newest modules, with STS-131 arriving in April 2010. The crew also saw the arrival of the Rasvet Russian docking module aboard STS-132, launched on May 14, 2010. The focal point of the Expedition 23 mission emblem is Earth with the ISS shown swooshing around it. The crew used Roman numerals as a design feature to allow display of the national flags of the crew.

EXPEDITION 24

DATE June 1–September 25, 2010
DOCKING April 4, 2010
UNDOCKING September 25, 2010
VEHICLES *Soyuz TMA-18*, *Soyuz TMA-19* (launch and return)
CREW Aleksandr Skvortsov, Mikhail Korniyenko, Tracy Caldwell Dyson, Douglas Wheelock, Shannon Walker, Fyodor Yurchikhin

This expedition experienced two significant mishaps. First, a cooling pump failed and cut power to parts of the ISS. Tracy Caldwell Dyson and Doug Wheelock worked exhaustively to repower the station. Second, the crew worked on a docking ring on the Mini-Research Module 2 Poisk, which delayed the return of *Soyuz TMA-18* to Earth. The mission patch placed the ISS in the foreground with the five rays of the Sun representing the international organizations supporting the ISS.

EXPEDITION 25

DATE September 25–November 26, 2010
DOCKING June 17, 2010
UNDOCKING November 25, 2010
VEHICLES Soyuz TMA-19, Soyuz TMA-01M (launch and return)
CREW Douglas Wheelock, Shannon Walker, Fyodor Yurchikhin, Scott Kelly, Aleksandr Kaleri, Oleg Skripochka

In addition to scientific experiments, the Expedition 25 crew met the *Progress M-08M* spacecraft on October 30, 2010, bringing 2.5 tons of cargo. This crew was also aboard during the 10th anniversary of occupancy on the ISS. The crew said that its mission patch, rich in symbolism, highlighted the Greek letter alpha as a symbol of the ISS in furthering space exploration. The overall patch form was a Greek letter omega, to recognize the role of the space shuttle in creating the ISS.

EXPEDITION 26

DATE November 26, 2010–March 16, 2011
DOCKING October 9, 2010
UNDOCKING May 23, 2011
VEHICLES Soyuz TMA-01M, Soyuz TMA-20 (launch and return)
CREW Scott Kelly, Aleksandr Kaleri, Oleg Skripochka, Dmitri Kondratyev, Catherine Coleman, Paolo Nespoli

Among other highlights, the Expedition 26 crew met the crew of STS-133 on February 26, 2011, and was resupplied by the ESA's *Johannes Kepler* resupply craft, which arrived on February 24. The crew's mission patch shows the resupply vehicle. The two stars symbolized two Soyuz capsules carrying the six members of the crew while the prominence of the ISS, according to the crew, acknowledged the efforts of the entire team.

The Japanese Experiment Module, Kibo, has a porch on the ISS for experimentation in space, shown here. Research on Kibo is focused on biotechnology and physical research.

CHAPTER 10
ISS EXPEDITIONS 2011–2020 »

ISS EXPEDITIONS 2011–2020

The second decade of operations on the ISS witnessed the transition from construction to completion and its use as a world-class scientific research laboratory for all manner of microgravity experiments. Crucially, this is an environment that cannot be duplicated on Earth. Here, gravity distorts the shape of crystalline structures, limiting their size and full three-dimensional structure. This research covered many areas. Biomedical research found application for Earth's population, such as knowledge about osteoporosis, dementia, and the circulatory system. It also facilitated groundbreaking research on lymph tissue, pivotal in the study of the human immunodeficiency virus (HIV).

Scientific knowledge about the human body—especially how it responds to long-duration spaceflight—gained a boost from data gathered on the physical changes the crews experienced. Materials science research has expanded human understanding of how various materials are affected by microgravity. Finally, the ISS has provided a platform for space science research into the origins and evolution of the Sun and the planets.

In the post–space shuttle era, the expeditions to the ISS were flown to the station at two separate times on two separate Soyuz capsules, three persons each. Take Expedition 27 as an example. Dmitri Kondratyev, Catherine Coleman, and Paolo Nespoli were already aboard the ISS and transferred from Expedition 26. They were joined later by Andrei Borisenko, Aleksandr Samokutyayev, and Ron Garan for the start of Expedition 27. These latter three crew members then transferred to Expedition 28 at the end of Expedition 27, and were joined by three additional crew members. This rotational approach has generally remained the case.

By the beginning of the 2020s, the ISS was slated to be operational through at least the latter part of the decade, offering nearly 30 years of operations for its scientific investigations. By early 2020, the total number of expedition crews stood at 61.

1]
NASA astronauts Scott Kelly (left) and Terry Virts (right), members of ISS Expedition 43, tend to experiments in the Japanese Experiment Module.

2]
The *Soyuz TMA-20* spacecraft departs from the ISS on May 23, 2011. Onboard are three members of Expedition 27, Dmitry Kondratyev, Catherine Coleman, and Paolo Nespoli.

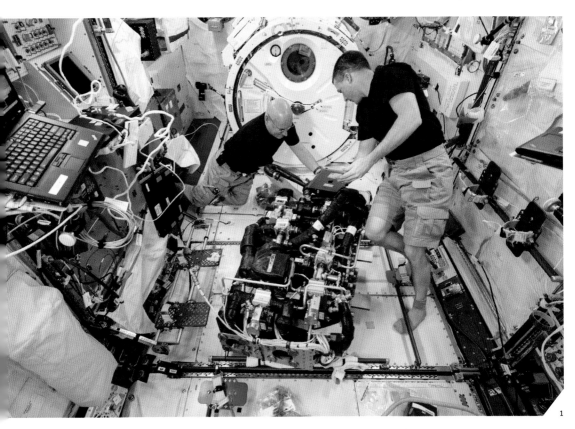

EXPEDITION 27

DATE March 16–May 23, 2011

VEHICLES *Soyuz TMA-20*, *Soyuz TMA-21* (launch and return)

CREW Dmitri Kondratyev, Catherine "Cady" Coleman, Paolo Nespoli, Andrei Borisenko, Aleksandr Samokutyayev, Ron Garan

A highlight of this expedition was the final rendezvous with the ISS of NASA's space shuttle *Endeavour*, on its last mission, STS-134. The central feature of the Expedition 27 patch was the ISS, showing the Alpha Magnetic Spectrometer, and two resupply vehicles docked at each end.

EXPEDITION 28

DATE May 23–September 15, 2011

VEHICLES *Soyuz TMA-21*, *Soyuz TMA-02M* (launch and return)

CREW Andrei Borisenko, Aleksandr Samokutyayev, Ron Garan, Sergey Volkov, Mike Fossum, Satoshi Furukawa

The highlight of this expedition was the docking of STS-135 with the ISS. Like most other mission patches, Expedition 28 features the ISS circling Earth, but shows the Moon to call attention to the possibility of a return. Since Expedition 28 took place during the 50th anniversaries of the first human in space (Russian cosmonaut Yuri Gagarin on April 12, 1961), and the first American in space (astronaut Alan Shepard on May 5, 1961), the patch has "50 Years" emblazoned on it with both Gagarin's and Shepard's names.

EXPEDITION 29

DATE September 15–November 21, 2011

VEHICLES *Soyuz TMA-02M*, *Soyuz TMA-22* (launch and return)

CREW Mike Fossum, Sergey Volkov, Satoshi Furukawa, Anton Shkaplerov, Anatoli Ivanishin, Dan Burbank

With the completion of the ISS, the crew of Expedition 29 focused on research. They envisioned paving the way for future missions beyond low Earth orbit, and inspiring young people to join in this adventure. The Expedition 29 patch features Captain James Cook's ship, the *Endeavour*, with the ISS following its path. The crew commented that the "ISS sails a stardust trail—following the spirit of the *Endeavour* sailing toward the dark unknown and new discoveries—it enlightens Earth below."

EXPEDITION 30

DATE November 21, 2011–April 27, 2012

VEHICLES *Soyuz TMA-22*, *Soyuz TMA-03M* (launch and return)

CREW Dan Burbank, Anton Shkaplerov, Anatoli Ivanishin, Oleg Kononenko, André Kuipers, Donald Pettit

This expedition saw the arrival of the ESA's third Automated Transfer Vehicle, *Edoardo Amaldi*, which docked successfully with the ISS on March 28, 2012, carrying over 14,500 pounds of propellants, supplies, and equipment. In addition, the Russian *Progress M-15M* resupply spacecraft docked on April 22, 2012. In recognition of the completion of the ISS, this mission patch depicts a fully assembled station. The crew wanted to emphasize the Earth at night, showing lights from space.

EXPEDITION 31

DATE April 27–July 1, 2012
VEHICLES *Soyuz TMA-03M, Soyuz TMA-04M* (launch and return)
CREW Oleg Kononenko, André Kuipers, Donald Pettit, Joseph Acaba, Gennady Padalka, Sergei Revin

The most significant event of Expedition 31 was SpaceX's *Dragon* spacecraft test rendezvous with the ISS as part of NASA's Commercial Orbital Transportation Services program. *Dragon* launched on May 22, docked with the ISS on May 25, and returned to Earth on May 31. The shape of this mission patch represents the Milky Way. The black background, said the crew, "symbolizes the research into dark matter." It also shows the Earth, Moon, Mars, and asteroids, which the crew considered the focus of future space exploration.

EXPEDITION 32

DATE July 1–September 16, 2012
VEHICLES *Soyuz TMA-04M, Soyuz TMA-05M* (launch and return)
CREW Gennady Padalka, Joseph Acaba, Sergei Revin, Sunita Williams, Yuri Malenchenko, Akihiko Hoshide

This expedition emphasized research using the JAXA Kibo laboratory. The Expedition 32 crew emphasized that the arch shape on its mission patch symbolized the "doorway" to future space exploration possibilities. The ISS with the torch above called attention to the pursuit of knowledge. The astronaut emblem also takes center stage on this mission patch.

EXPEDITION 33

DATE September 16–November 18, 2012
VEHICLES *Soyuz TMA-05M*, *Soyuz TMA-06M* (launch and return)
CREW Sunita Williams, Yuri Malenchenko, Akihiko Hoshide, Kevin Ford, Oleg Novitskiy, Evgeny Tarelkin

The Expedition 33 crew successfully experimented with the delay-tolerant networking protocol and managed to control a Lego™ robot on Earth from space. With the flags of Japan, Russia, and the U.S., this mission patch incorporates five white stars representing the partners participating in the ISS program—Canada, the ESA, Japan, Russia, and the U.S. Expedition 33 focused on biology and biotechnology, and other sciences.

EXPEDITION 34

DATE November 18, 2012–March 15, 2013
VEHICLES *Soyuz TMA-06M*, *Soyuz TMA-07M* (launch and return)
CREW Kevin Ford, Oleg Novitskiy, Evgeny Tarelkin, Thomas Marshburn, Chris Hadfield, Roman Romanenko

Life science experimentation received major emphasis on Expedition 34, especially investigations of the human cardiovascular system in space, studies on fish and their sensation of gravity, and the impact of solar radiation on life-forms. In addition to the depiction of the ISS, the Expedition 34 crew said of their patch: "Inside in gold is a craft symbolizing future extra-terrestrial landers that will someday open other celestial destinations to human exploration."

EXPEDITION 35

DATE March 15, 2013–May 14, 2013

VEHICLES *Soyuz TMA-07M*, *Soyuz TMA-08M* (launch and return)

CREW Chris Hadfield, Thomas Marshburn, Roman Romanenko, Christopher Cassidy, Pavel Vinogradov, Aleksandr Misurkin

During Expedition 35, the SpaceX CRS-2 mission successfully delivered supplies to the station and returned some cargo from space. In addition, ISS commander Chris Hadfield gained celebrity by posting the "first music video recorded in space" on YouTube, a rendition of David Bowie's 1969 song "Space Oddity." This patch featured a less busy design than most, on which the Expedition 35 crew commented: "The arc of the Earth's horizon with the Sun's arrows of light imply a bow shooting the imagination to Mars and the cosmos where our species may one day thrive."

EXPEDITION 36

DATE May 14, 2013–September 10, 2013

VEHICLES *Soyuz TMA-08M*, *Soyuz TMA-09M* (launch and return)

CREW Pavel Vinogradov, Christopher Cassidy, Aleksandr Misurkin, Karen Nyberg, Fyodor Yurchikhin, Luca Parmitano

During a space walk on July 16, 2013, Luca Parmitano reported that water was steadily leaking into his helmet. Flight controllers aborted the EVA, and Parmitano made his way safely back to the airlock. The dynamic design of the Expedition 36 mission patch portrays the ISS's solar arrays. The slanted angles, according to the crew, denoted a kinetic energy leading from the Earth in the lower right to the upper left tip of the triangular patch.

EXPEDITION 37

DATE September 10–November 10, 2013

VEHICLES *Soyuz TMA-09M, Soyuz TMA-10M* (launch and return)

CREW Fyodor Yurchikhin, Karen Nyberg, Luca Parmitano, Oleg Kotov, Sergey Ryazansky, Michael Hopkins

Expedition 37 focused on biomedical research, with two members of the crew specialists in the field. Using Leonardo da Vinci's *Vitruvian Man* on its mission patch, the Expedition 37 crew viewed it as a perfect symbol for the symmetry of the ISS and its multifaceted research.

EXPEDITION 38

DATE November 10, 2013–March 10, 2014

VEHICLES *Soyuz TMA-10M, Soyuz TMA-11M* (launch and return)

CREW Oleg Kotov, Sergey Ryazansky, Michael Hopkins, Koichi Wakata, Richard Mastracchio, Mikhail Tyurin

Expedition 38 included a variety of experiments, ranging from protein crystal growth studies to biological studies of plant seedlings. A key research area for this expedition was human health for long-duration space travel, as NASA and Roscosmos prepared for two crew members to spend a full year aboard the space station in 2015. The crew wanted the mission patch to offer a scenario for exploration beyond low Earth orbit, so the flowing mission numbers that wrap around Earth point toward the Moon and Mars.

EXPEDITION 39

DATE March 11–May 13, 2014

VEHICLES *Soyuz TMA-11M*, *Soyuz TMA-12M* (launch and return)

CREW Koichi Wakata, Richard Mastracchio, Mikhail Tyurin, Aleksandr Skvortsov, Oleg Artemyev, Steven Swanson

Expedition 39 was the first time the ISS was commanded by a Japanese astronaut, Koichi Wakata. The elegant design of the Expedition 39 mission patch offers a stylized portrayal of a Soyuz spacecraft forging the future. As the transport vehicle for the crew members to and from the station, they believed it deserved special recognition in the emblem.

EXPEDITION 40

DATE May 13–September 10, 2014

VEHICLES *Soyuz TMA-12M*, *Soyuz TMA-13M* (launch and return)

CREW Steven Swanson, Aleksandr Skvortsov, Oleg Artemyev, Gregory Wiseman, Maksim Surayev, Alexander Gerst

Expedition 40 emphasized research on Earth, remote sensing, an assessment of human behavior and performance, and studies of animal biology and bone and muscle physiology. Strikingly ornate, the Expedition 40 crew wrote about their mission patch: "The reliable and proven Soyuz, our ride to the International Space Station (ISS), is a part of the past, present, and future . . . The sun on Earth's horizon represents the new achievements and technologies that will come about due to our continued effort in space exploration."

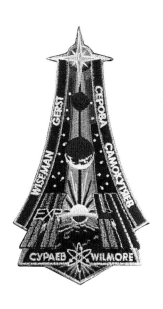

EXPEDITION 41

DATE September 10–November 10, 2014

VEHICLES *Soyuz TMA-13M*, *Soyuz TMA-14M* (launch and return)

CREW Maksim Surayev, Gregory Wiseman, Alexander Gerst, Aleksandr Samokutyayev, Yelena Serova, Barry Wilmore

The Expedition 41 crew wrote, "Portraying the road of human exploration into our vastly unknown universe, all elements of the Expedition 41 patch build from the foundation, our Earth, to the stars beyond our solar system. The focus of our six-month expedition to the ISS is Earth and its inhabitants as well as a scientific look out into our universe."

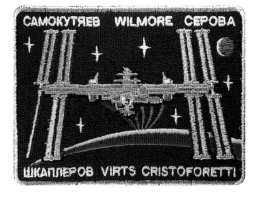

EXPEDITION 42

DATE November 10, 2014–March 11, 2015

VEHICLES *Soyuz TMA-14M*, *Soyuz TMA-15M* (launch and return)

CREW Barry Wilmore, Aleksandr Samokutyayev, Yelena Serova, Anton Shkaplerov, Samantha Cristoforetti, Terry Virts

Expedition 42 worked on reconfiguration of the ISS to allow commercial space taxis to dock at the station, requiring multiple space walks to prepare the Harmony node. Paying homage to the Expedition 1 mission patch, the Expedition 42 mission patch portrays the ISS orbiting Earth with its solar array wings spread wide. The crew believed showing the solar arrays capturing the Sun portrayed their fundamental mission, to use the capability of the space station to undertake scientific research.

EXPEDITION 43

DATE March 11–June 11, 2015

VEHICLES *Soyuz TMA-15M*, *Soyuz TMA-16M* (launch and return)

CREW Terry Virts, Anton Shkaplerov, Samantha Cristoforetti, Gennady Padalka, Mikhail Korniyenko, Scott Kelly

The Expedition 43 crew spent an extra "bonus month" on board while an investigation of a failure of a Russian *Progress M-27M* cargo mission took place. Additionally, on June 8, 2015, the ISS adjusted its orbit to move to a safe distance from a piece of orbital space debris. A relatively simple design, this mission patch graphically emphasized the expedition number. The hexagon shape also represented the six crew members onboard the orbital outpost.

EXPEDITION 44

DATE June 11–September 11, 2015

VEHICLES *Soyuz TMA-16M*, *Soyuz TMA-17M* (launch and return), *Soyuz TMA-18M* (return)

CREW Gennady Padalka, Mikhail Korniyenko, Scott Kelly, Oleg Kononenko, Kimiya Yui, Kjell Lindgren

Two members of the Expedition 44 crew became the first Americans ever to eat food grown entirely in space. Scott Kelly and Mikhail Korniyenko also began a yearlong stay on the ISS. An oval design with the Earth circling the Sun and the ISS at the bottom center, this mission patch emphasizes the duration of their effort. The 12 Earths represented the planet's position around the Sun over the course of that year. Four of the Earths were silhouetted in sunlight, representing the four-month duration of Expedition 44.

EXPEDITION 45

DATE September 11–December 11, 2015

VEHICLES Soyuz TMA-16M, Soyuz TMA-17M (launch and return), Soyuz TMA-18M (return)

CREW Scott Kelly, Oleg Kononenko, Mikhail Korniyenko, Kimiya Yui, Kjell Lindgren, Sergey Volkov

Scott Kelly and Mikhail Korniyenko transferred from Expedition 44 as part of their yearlong stay aboard the ISS. Kelly, Korniyenko, and Sergey Volkov also transferred to the crew of Expedition 46. This mission patch showed the ISS circling Earth with the Moon and Mars in the distance. The crew incorporated a book into the bottom of the patch to reflect "the flow of knowledge, hard work, sacrifice, and innovation that makes human spaceflight possible."

EXPEDITION 46

DATE December 11, 2015–March 1, 2016

VEHICLES Soyuz TMA-16M (launch), Soyuz TMA-18M, Soyuz TMA-19M (launch and return)

CREW Scott Kelly, Mikhail Korniyenko, Sergey Volkov, Yuri Malenchenko, Timothy Peake, Timothy Kopra

The *Progress MS-1* resupply mission launched to the ISS on December 21. To prepare for its arrival, Commander Scott Kelly and Timothy Kopra performed a contingency extravehicular activity and successfully repaired the Mobile Base System. A simple triangle shows only a bold Expedition 46 number, the names of the crew, and flags at the top.

EXPEDITION 47

DATE March 2–June 18, 2016
VEHICLES *Soyuz TMA-16M* (launch), *Soyuz TMA-18M*, *Soyuz TMA-19M* (launch and return)
CREW Yuri Malenchenko, Timothy Peake, Timothy Kopra, Aleksey Ovchinin, Oleg Skripochka, Jeffrey Williams

Launched on April 8, 2016, the SpaceX CRS-8 mission carried the Bigelow Expandable Activity Module to the ISS for two years of in-orbit habitat qualification. Like a fisheye lens, this mission patch depicts the ISS with the Moon above as a future destination for space exploration. The Expedition 47 crew names are written in their native languages around the outside of the patch.

EXPEDITION 48

DATE June 18–September 7, 2016
VEHICLES *Soyuz TMA-20M*, *Soyuz MS-01* (launch and return)
CREW Jeffrey Williams, Aleksey Ovchinin, Oleg Skripochka, Kathleen Rubins, Anatoli Ivanishin, Takuya Onishi

Conceptual rather than realistic, the Expedition 48 mission patch points toward future space exploration. As stated by NASA, "The elements of the crew patch include ISS solar arrays illuminated by the setting Sun, the Earth's horizon at sunset, the Moon, and stars. The simple portrayal of the unique vantage point signifies the incremental contribution of a single international expedition off the planet to the larger endeavor of human space exploration and discovery."

EXPEDITION 49

DATE September 7–October 30, 2016

VEHICLES *Soyuz MS-01*, *Soyuz MS-02* (launch and return)

CREW Anatoli Ivanishin, Kathleen Rubins, Takuya Onishi, Robert Kimbrough, Andrei Borisenko, Sergey Ryzhikov

Expedition 49 continued to focus on biomedical research. With the aurora borealis dominating the Expedition 49 mission patch, the ISS is also depicted in silhouette at the bottom. It was designed at the Johnson Space Center by graphics artist Cindy Bush.

EXPEDITION 50

DATE October 28, 2016–April 10, 2017

VEHICLES *Soyuz MS-02*, *Soyuz MS-03* (launch and return), *Soyuz MS-04* (return)

CREW Robert Kimbrough, Andrei Borisenko, Sergey Ryzhikov, Peggy Whitson, Oleg Novitskiy, Thomas Pesquet

Reminiscent of the work of graphic artist Robert Rauschenberg, the Expedition 50 mission patch featured Earth, the Moon, and the ISS with an oversized expedition number. Designed by Sean Collins, this boxlike graphic design represented a simple story; the ISS serves as a stepping-stone between the Earth and the Moon.

EXPEDITION 51

DATE April 10–June 2, 2017

VEHICLES *Soyuz MS-03*, *Soyuz MS-04* (launch and return)

CREW Peggy Whitson, Oleg Novitskiy, Thomas Pesquet, Fyodor Yurchikhin, Jack Fischer

The principal designer for this unique shield design was Johnson Space Center artist Sean Collins, who took the ideas of the crew, especially Peggy Whitson, into account in designing it. Whitson said, "From as early as the 11th century, coats of arms have been used as emblems representing groups as small as families to as large as countries. The Expedition 51 patch is designed as a modernized international coat of arms."

EXPEDITION 52

DATE June 2–September 2, 2017

VEHICLES *Soyuz MS-03*, *Soyuz MS-04* (launch and return), *Soyuz MS-05* (launch)

CREW Fyodor Yurchikhin, Jack Fischer, Peggy Whitson, Sergey Ryazansky, Randolph Bresnik, Paolo Nespoli

After a succession of simple and elegant designs, the Expedition 52 crew took a different approach with a colorful and busy mission patch. Designed by Dutch artist Luc van den Abeelen, together with cosmonaut Fyodor Yurchikhin, this patch shows Earth in the center with the ISS and *Sputnik* in orbit.

EXPEDITION 53

DATE September 12–December 14, 2017

VEHICLES Soyuz MS-05, Soyuz MS-06 (launch and return)

CREW Randolph Bresnik, Paolo Nespoli, Sergey Ryazansky, Alexandr Misurkin, Mark Vande Hei, Joseph Acaba

With the Earth forming part of the Expedition 53 number on this patch, the ISS circles and points toward Mars as an eventual destination. Designed by Kennedy Space Center artists under the direction of Expedition 53 commander Randolph Bresnik, this mission patch bears a striking resemblance to the STS-129 mission patch (see page 154), the design of which was also overseen by Bresnik.

EXPEDITION 54

DATE December 17, 2017–February 27, 2018

VEHICLES Soyuz MS-06, Soyuz MS-07 (launch and return)

CREW Alexander Misurkin, Mark Vande Hei, Joseph Acaba, Anton Shkaplerov, Scott Tingle, Norishige Kanai

This crew performed three space walks to repair elements of the ISS, totaling more than 21 hours. Reminiscent of imagery from the *Star Wars* films, the Expedition 54 mission patch had both a stylized ISS and the Moon in its design. This insignia was created by Sean Collins.

EXPEDITION 55

DATE February 27–June 3, 2018

VEHICLES *Soyuz MS-07*, *Soyuz MS-08* (launch and return)

CREW Anton Shkaplerov, Scott Tingle, Norishige Kanai, Andrew Feustel, Oleg Artemyev, Richard Arnold

The astronaut emblem takes center stage in the Expedition 55 mission patch. The six crew members of Expedition 55 are from three different countries—Japan, Russia, and the U.S.—and the three flags are depicted in the astronaut emblem at the center.

EXPEDITION 56

DATE June 3–October 4, 2018

VEHICLES *Soyuz MS-08*, *Soyuz MS-09* (launch and return)

CREW Andrew Feustel, Oleg Artemyev, Richard Arnold, Alexander Gerst, Serena Auñón-Chancellor, Sergey Prokopyev

In a departure from earlier expeditions, a change of command took place on June 1, 2018, with the official beginning of Expedition 56 on June 3. On August 29, 2018, mission controllers detected a leak through a drop in air pressure. Astronauts then discovered a 2mm hole in the *Soyuz MS-09* spacecraft. They initially repaired the hole with tape, followed by a permanent repair with gauze and epoxy. With a Soyuz launch on the left and the ISS on the right, the Expedition 56 mission patch is dominated by a dove in the center.

EXPEDITION 57

DATE October 4–December 20, 2018

VEHICLES *Soyuz MS-09*, *Soyuz MS-11* (launch and return), *Soyuz MS-10* (failed)

CREW Alexander Gerst, Serena Auñón-Chancellor, Sergey Prokopyev, Oleg Kononenko, Anne McClain, David Saint-Jacques

While three members of the Expedition 56 crew transitioned to Expedition 57, Aleksey Ovchinin and Nick Hague were to join them on October 11, 2018, on *Soyuz MS-10*, which aborted during launch due to a booster failure. The crew landed safely after a ballistic descent. Since the initial Expedition 57 crew needed to depart by mid-December 2018 in *Soyuz MS-09* due to the capsule's limited on-orbit life span, space agency officials decided to fly another crew to the ISS on *Soyuz MS-11*, launched on December 3, 2018.

EXPEDITION 58

DATE December 20, 2018–March 15, 2019

VEHICLES *Soyuz MS-11* (launch), *Soyuz MS-11* (return)

CREW Oleg Kononenko, Anne McClain, David Saint-Jacques

The initial Expedition 57 crew departed on December 20, and Expedition 58 started as a three-person increment. Because of the failure of *Soyuz MS-10*, Expedition 58 had only three crew members. The Earth and the ISS are featured in the Expedition 58 mission patch. Central to the patch is the compass rose, which the crew suggested was a symbol of all exploration. "The stars on the Expedition 58 patch are their families," they added, "one star for each member. They shine on as a beacon of strength and a guiding light home."

EXPEDITION 59

DATE March 15–June 24, 2019

VEHICLES Soyuz MS-11 and Soyuz MS-12 (launch and return)

CREW Oleg Kononenko, Anne McClain, David Saint-Jacques, Aleksey Ovchinin, Nick Hague, Christina Koch

Going back to a six-person crew, researchers on Expedition 59 conducted experiments on tissue chips in the microgravity environment to study the effects of aging and disease. The mission patch played with the design of an atom to depict the ISS as a nucleus while the nations involved in the effort were orbiting electrons. The crew claimed that, like electrons in an atom, international cooperation is the basic stabilizing force that enables large-scale space exploration. The patch also focused on the ISS as a microgravity science laboratory.

EXPEDITION 60

DATE June 24–October 3, 2019

VEHICLES Soyuz MS-12 (launch and return) and Soyuz MS-13 (launch)

CREW Aleksey Ovchinin, Nick Hague, Christina Koch, Aleksandr Skvortsov, Luca Parmitano, Andrew Morgan

Expedition 60 took place during the 50th anniversary of the first Moon landing. This mission patch commemorates the event with a constellation of three stars with the Moon superimposed forming the letter L, the Roman numeral for 50. The Moon is depicted as a waxing crescent, as it was on July 20, 1969. The familiar silhouette of the ISS is visible, flying across the night sky.

EXPEDITION 61

DATE October 3, 2019–February 6, 2020
VEHICLES *Soyuz MS-12*, *Soyuz MS-15* (launch), *Soyuz MS-13* (launch and return)
CREW Luca Parmitano, Aleksandr Skvortsov, Andrew Morgan, Christina Koch, Oleg Skripochka, Jessica Meir

Expedition 61 was the last mission completed at the time of writing. The crew conducted nine space walks, more than any other crew in the history of the ISS, totaling more than 54 hours repairing elements of the station. For their mission patch, the Expedition 61 crew used a stylized version of the ISS. NASA commented: "The overall view is from an approaching vehicle in pursuit of the space station. The Sun is the most prominent element in the patch as the source of energy and life for the Earth, the station, and the solar system. Fifteen of the Sun's rays represent the 15 original partner members of the space station program, while the 16th ray represents an open invitation for continued collaboration with new partners."

CONCLUSION

NASA's human spaceflight efforts would be viewed quite differently had there never been mission patches signifying the efforts of the crews and their ground support over the decades. They are a traditional part of the public involvement in human spaceflight. Superficially, they do not amount to much: small and colorful stitched fabric that may be sewn or ironed onto clothing, traded, or displayed on walls. Collecting the patches, studying their nuances, and discussing them within a community of enthusiasts remains a hobby of choice for thousands.

Why do those collecting them do so? Some have financial gain as their motive, and there is not much this book could tell them that they do not already know. Indeed, I learned what I know about mission patches from those collectors, and I am grateful for their diligence and willingness to educate others. For others, and I place myself in that category, the interest in mission patches began with human space exploration itself. Mission patches provide an insight into the larger efforts toward discovery and exploration. They enhance our passion for spaceflight, while also giving us a sense of ownership. We might never have met the crew of STS-120, but when we have the mission patch that they proudly wore on their flight clothing, we feel a tangible connection to those astronauts and their profoundly important work.

The mission patches that I have acquired over the years bring me back to my youth. They remind me of when I wrote to NASA during the Space Race and asked for information about the space program. Virtually every response I received included one or more mission insignia of some type. In every case it made me want to learn more about NASA's efforts, the individual missions signified by those mission patches, and the prospects for future space exploration.

The patches portrayed in these pages show us how, from the very beginning of the Space Age, graphic design, colored borders, various artistic representations, and sometimes psychedelic layouts have evolved. What they depict provides a unique insight into the accomplishments of astronauts, while occasionally providing moments of comic relief and sublime delight. At least they do for me. I hope the same is true for you.

1] Fred Haise served on *Apollo 13*. During this vist to the Johnson Space Center in 2007, he told engineers not to forget the lessons of that mission, mainly to make a space vessel hardy enough to handle surprises.

2] In this poster produced for NASA's Space Flight Awareness Office, the ISS Expedition 45 crew posed as lightsaber-wielding Jedi knights from the *Star Wars* films for their official mission portrait. The mission patch is in the center.

3] At the end of the space shuttle program, on August 4, 2011, NASA's Space Flight Awareness Office produced this poster containing the mission patches for all of the shuttle flights.

FURTHER READING

Key Books on Human Spaceflight

Brinkley, Douglas. *American Moonshot: John F. Kennedy and the Great Space Race.* New York: Harper, 2019. This is a vivid account focused on the relationship of JFK to the space efforts at NASA. It is a richly textured account, much better for political history than most other works on the Apollo program.

Burrows, William E. *This New Ocean: The Story of the First Space Age.* New York: Random House, 1998. A strong overview of the history of the Space Age from *Sputnik* to 1998.

Compton, W. David, and Charles D. Benson. *Living and Working in Space: A History of Skylab.* Washington, DC: NASA Special Publication-4208, 1983. The official NASA history of *Skylab*.

—*Where No Man Has Gone Before: A History of Apollo Lunar Exploration Missions.* Washington, DC: NASA Special Publication-4214, 1989. This clearly written account traces the ways in which science went to the Moon.

Ezell, Edward Clinton, and Linda Neuman Ezell. *The Partnership: A History of the Apollo-Soyuz Test Project.* Washington, DC: NASA Special Publication-4209, 1978. A detailed study of the effort by the U.S. and the Soviet Union in the mid-1970s to conduct a joint human spaceflight.

Hacker, Barton C., and James M. Grimwood. *On Shoulders of Titans: A History of Project Gemini.* Washington, DC: NASA Special Publication-4203, 1977. The official history of the Gemini project conducted by NASA in the mid-1960s.

Hersch, Matthew M. *Inventing the American Astronaut.* New York: Palgrave Macmillan, 2012. An important discussion of the early astronaut program.

Jenkins, Dennis R. *Space Shuttle: The History of the National Space Transportation System.* 3 Vols. Cape Canaveral, FL: Dennis R. Jenkins, 2016, 5th Edition. Perhaps the best technical history, presenting an overview of the space shuttle and its development and use.

Launius, Roger D. *Apollo's Legacy: Perspectives on the Moon Landings.* Washington, DC: Smithsonian Books, 2019. An analysis of the meaning of the Apollo program after 50 years.

—*Reaching for the Moon: A Short History of the Space Race.* New Haven, CT: Yale University Press, 2019. A concise history of the space race between the U.S. and the Soviet Union.

Logsdon, John M. *After Apollo? Richard Nixon and the American Space Program.* New York: Palgrave Macmillan, 2015. On July 20, 1969, Neil Armstrong took "one small step for a man, one giant leap for mankind." The success of the *Apollo 11* mission satisfied the goal that had been set by President John F. Kennedy just over eight years earlier. It also raised the question "What do you do next after landing on the Moon?" It fell to President Richard M. Nixon to answer this question. This book traces, in detail, how Nixon and his associates went about developing their response.

—General Editor. *Exploring the Unknown: Selected Documents in the History of the U.S. Civil Space Program.* 6 Vols. Washington, DC: NASA Special Publication-4407, 1995–2004. An essential reference, these volumes print more than 350 key documents in space policy and its development throughout the 20th century.

—*John F. Kennedy and the Race to the Moon.* New York: Palgrave Macmillan, 2010. This study, based on extensive research in primary documents and archival interviews, is the definitive examination of John F. Kennedy's role in sending Americans to the Moon. Among other revelations, the author finds that following the Cuban Missile Crisis in 1962 Kennedy pursued an effort to turn Apollo into a cooperative program with the Soviet Union.

McCurdy, Howard E. *Space and the American Imagination.* Washington, DC: Smithsonian Institution Press, 1997. A significant analysis of the relationship between popular culture and public policy.

McDougall, Walter A. *The Heavens and the Earth: A Political History of the Space Age.* New York: Basic Books, 1985. Reprint edition, Baltimore, MD: Johns Hopkins University Press, 1997. This Pulitzer Prize–winning book analyzes the Space Race in the 1960s. The author argues that Apollo prompted the space program to stress engineering over science, competition over cooperation, civilian over military management, and international prestige over practical applications.

Muir-Harmony, Teasel E. *Apollo to the Moon: A History in 50 Objects.* Washington, DC: National Geographic, 2019. An illustrated history of key artifacts in the history of Apollo.

Murray, Charles A., and Catherine Bly Cox. *Apollo: The Race to the Moon.* New York: Simon and Schuster, 1989. Reprint edition, Burkittsville, MD: South Mountain Books, 2004. Perhaps the best general account of the lunar program, this history uses interviews and documents to reconstruct the stories of the people who participated in Apollo.

Neal, Valerie. *Spaceflight in the Shuttle Era and Beyond: Redefining Humanity's Purpose in Space.* New Haven, CT: Yale University Press, 2017. An important interpretation of the space shuttle's meaning in American culture.

Neufeld, Michael J. *Von Braun: Dreamer of Space, Engineer of War.* New York: Alfred A. Knopf, 2007. By far the finest biography of the German rocketeer émigré.

Scott, David Meerman, and Richard Jurek. *Marketing the Moon: The Selling of the Apollo Lunar Program.* Cambridge, MA: MIT Press, 2014. An illustrated work on the sophisticated efforts by NASA and its many contractors to market the facts about space travel—through press releases, bylined articles, lavishly detailed background materials, and fully produced radio and television features—rather than push an agenda.

Stone, Robert, and Alan Andres. *Chasing the Moon: The People, the Politics, and the Promise That Launched America into the Space Age.* New York: Ballantine Books, 2019. A companion to the PBS documentary, this book offers new perspectives on the stories of the Moon landing.

Swenson, Loyd S., Jr., James M. Grimwood, and Charles C. Alexander. *This New Ocean: A History of Project Mercury.* Washington, DC: NASA Special Publication-4201, 1966. The official history of Project Mercury, this book is based on extensive research and interviews.

Vaughan, Diane. *The Challenger Launch Decision: Risky Technology, Culture, and Deviance at NASA.* Chicago: University of Chicago Press, 1996. The first thorough scholarly study of the events leading to the fateful decision to launch *Challenger* in January 1986.

Weitekamp, Margaret A. *Right Stuff, Wrong Sex: America's First Women in Space Program.* Baltimore, MD: Johns Hopkins University Press, 2004. On June 17, 1963, Soviet cosmonaut Valentina Tereshkova became the first woman in space. Unlike every previous milestone in the "space race," however, this event did not spur NASA to put an American woman into orbit. There were suitable candidates: two years earlier, thirteen female pilots recruited by the private Woman in Space program had passed a strenuous physical exam and were ready for another stage of astronaut testing. Yet American women did not escape Earth's orbit for another thirty years.

Wolfe, Tom. *The Right Stuff.* New York: Farrar, Straus and Giroux, 1979. An outstanding journalistic account of the first years of spaceflight, essentially Project Mercury, focusing on the Mercury Seven astronauts.

Key Books and Articles on Mission Patches

Glushko, Alexander. *Design for Space: Soviet and Russian Mission Patches.* Berlin, Germany: DOM Publishing, 2016. Covering the period from the beginning of the Cold War to the first crew on the International Space Station, this book documents the almost 250 mission patches worn by Soviet and Russian cosmonauts since 1963.

Hengeveld, Ed. "The Apollo Emblems of Artist Al Stevens." *Spaceflight*, June 2008, pp. 220–225. Available online at http://www.collectspace.com/news/news-052008a.html. An account of the art of Al Stevens, designer of several mission patches.

Kaplan, Judith, and Robert Muniz. *Space Patches: From Mercury to the Space Shuttle.* New York: Sterling Publishing Co., 1986. Goes into depth on how NASA's human spaceflight patches came to fruition.

Kircher, Travis. "More Than Just a Merit Badge." *Ad Astra*, Nov/Dec 2000, pp. 23–25. Available online at http://www.collectspace.com/resources/patches_astronauts.html. An appreciation of the art that goes into mission patch design and its evolution.

Lattimer, Dick. *All We Did Was Fly to the Moon.* Gainesville, FL: The Whispering Eagle Press, 1985. An encyclopedic account of the missions and patches developed for the Mercury, Gemini, Apollo, Skylab, and ASTP programs.

Vogt, Gregory. *Space Mission Patches.* Minneapolis, MN: Millbrook Press, 2001. A book for younger readers depicting mission patches.

Key Websites for Mission Patches

abemblem.com An excellent site overseen by the company that has manufactured the majority of the mission patches used by NASA.

collectspace.com An indispensable source of information and interchange on space memorabilia, including mission patches, overseen by Robert Pearlman.

genedorr.com/patches/Intro.html A superb website overseen by Gene Dorr focusing on the intricacies of collecting mission patches for the "Golden Age" of spaceflight.

spacepatches.nl A website dedicated to collecting mission patches, overseen by Jacques Edwin van Oene and Erik van der Hoorn.

GLOSSARY OF TERMS

ACTS: Advanced Communications Technology Satellite.
Agena: The upper stage of a missile used as a target vehicle for rendezvous and docking during Project Gemini.
Apollo, Project: The third NASA human spaceflight program, ultimately undertaking a series of human landings on the Moon.
Apollo command and service module: The capsule and support systems used by the astronauts during Project Apollo.
ASI: Italian Space Agency (*Agenzia Spaziale Italiana*).
ASTP: Apollo-Soyuz Test Project—the 1975 cooperative flight by the United States and the Soviet Union.
ATLAS: Atmospheric Laboratory for Applications and Science.
CDRA: Carbon Dioxide Removal Assembly.
CNES: Centre National d'études Spatiales, the French government space agency.
CRISTA-SPAS: Flying Cryogenic Infrared Spectrometers and Telescopes for the Atmosphere-Shuttle Pallet Satellite.
CRS: Commercial Resupply Service.
CSA: Canadian Space Agency.
DARA: German Space Agency (Deutsches Zentrum für Luft- und Raumfahrt).
DoD: Department of Defense.
ESA: European Space Agency.
ET: External fuel tank on the space shuttle.
EURECA: European Retrievable Carrier.
EVA: Extravehicular activity, also called a space walk.
F-1: The main rocket engine used on the first stage of the Saturn V rocket.
Gemini, Project: The second NASA human spaceflight program, with piloted flights flown in 1965–1966.
GRO: Gamma Ray Observatory.
IML: International Microgravity Laboratory.
ISS: International Space Station.

JAXA: Japanese Space Exploration Agency.
JEM: Japanese Experiment Module.
JSC: Johnson Space Center.
LDEF: Long-Duration Exposure Facility.
LITE: Lidar In-Space Technology Experiment.
LMS: Life and Microgravity Spacelab.
LOX: Liquid oxygen, which was used as an oxidizer for the liquid hydrogen fuel used on several NASA rockets.
Mir: Soviet/Russian space station flown between 1986 and 2001.
MLPM: Multi-Purpose Logistics Module.
MSC: Manned Spacecraft Center.
Mercury, Project: The first NASA human spaceflight program, with human missions flown between 1961 and 1963.
Mercury Seven: The original astronauts chosen by NASA in 1959.
NASA: National Aeronautics and Space Administration.
NICMOS: Near Infrared Camera and Multi-Object Spectrometer.
OAST-Flyer: Office of Aeronautics and Space Technology-Flyer.
ORFEUS-SPAS: Orbiting and Retrievable Far and the Extreme Ultraviolet Spectrograph-Shuttle Pallet Satellite.
RSA: Russian Aviation and Space Agency, later Roscosmos.
Saturn V: The three-stage rocket that took astronauts to the Moon during Project Apollo.
Skylab: The orbital workshop where three astronauts crews worked on long duration spaceflight between 1973 and 1974.
SLS: Spacelab Life Sciences.
Soyuz: Soviet/Russian rocket and space capsule used for human space exploration missions.
Soyuz-MS: The latest version of the Soyuz spacecraft.
Soyuz TMA-M: Spacecraft used by the Russian Aviation and Space Agency for human spaceflight.

Spacelab: A science experiments module built by the European Space Agency that fit into the payload bay of the space shuttle.
Space shuttle: The longest NASA human spaceflight program, flying into space from 1981 to 2011.
SRL: Space Radar Laboratory.
SRB: Solid rocket boosters, used to launch the space shuttle.
SRTM: Shuttle Radar Topography Mapping mission.
SSBUV: Shuttle Solar Backscatter Ultraviolet spectrometer.
STIS: Space Telescope Imaging Spectrograph.
STS: Space Transportation System, an alternative term used to refer to the space shuttle program.
TDRSS: Tracking and Data Relay Satellite System.
TSS: Tethered Satellite System.
UARS: Upper Atmosphere Research Satellite.
USML: U.S. Microgravity Laboratory.

PICTURE CREDITS

All NASA mission patch images supplied by A-B Emblem unless stated otherwise below.

Cover: background: NASA; emblem: NASA/A-B Emblem
Back cover and spine: emblems: NASA/A-B Emblem
Front endpaper: background: NASA; emblem: NASA/A-B Emblem
Page 2: NASA
Page 7: top: NASA; bottom: NASA
Page 9: top left: NASA; top right: NASA; bottom: NASA
Page 11: top: NASA; bottom left: NASA; bottom right: NASA
Page 13: top left: NASA; top right: NASA; bottom: NASA
Page 15: top: NASA; bottom: NASA
Page 17: top: Wikimedia Commons; bottom: NASA and ESA
Page 18: NASA
Page 21: top left: NASA; top right: NASA; bottom: NASA
Page 22: left: NASA; right: NASA/A-B Emblem
Page 23: NASA
Page 24: NASA
Page 25: top: NASA; bottom: NASA
Page 26: NASA
Page 27: top: NASA; bottom: NASA
Page 28: NASA
Page 31: top: NASA; bottom: NASA
Page 32: NASA
Page 33: top left: NASA; top right: NASA; bottom: NASA
Page 34: right: NASA
Page 36: NASA
Page 37: top: NASA
Page 38: NASA
Page 40: top: NASA
Page 42: NASA
Page 45: top: NASA; bottom: NASA
Page 47: top left: NASA; top right: NASA; bottom: NASA
Page 49: top: NASA; bottom left: NASA; bottom right: NASA
Page 51: top: NASA; bottom left: NASA; bottom right: NASA
Page 53: bottom: NASA
Page 55: NASA
Page 57: NASA
Page 61: NASA
Page 68: NASA
Page 71: top: NASA; bottom left: NASA; bottom right: NASA
Page 73: top: NASA; bottom: NASA
Page 78: NASA
Page 81: top: NASA; bottom left: NASA; bottom right: NASA
Page 83: top left: NASA; top right: NASA (photo by Paul Weitz); bottom: NASA
Page 98: NASA
Page 101: top left: NASA; top right: NASA; bottom: NASA
Page 116: NASA
Page 119: top: NASA; bottom: NASA
Page 132: NASA (photo by Bill Ingalls)
Page 135: top: NASA; bottom: NASA
Page 158: NASA
Page 161: top: NASA (photo by Bill Ingalls); bottom: NASA
Page 163: top left: NASA; top right: NASA; bottom: NASA
Page 178: NASA
Page 181: top: NASA; bottom: NASA
Page 201: top left: NASA and Stennis Space Center; top right: NASA; bottom: Wikimedia Commons
Rear endpaper: NASA

All trademarks, trade names, and other product designations referred to herein are the property of their respective owners and are used solely for identification purposes. This book is a publication of The Bright Press, an imprint of The Quarto Group, and has not been authorized, licensed, approved, sponsored, or endorsed by any other person or entity. The publisher is not associated with any product, service, or vendor mentioned in this book. While every effort has been made to credit contributors, The Bright Press would like to apologize should there have been any omissions or errors, and would be pleased to make the appropriate correction for future editions.

INDEX

Page numbers in *italic* type refer to mission photographs.

A

A-B Emblem 16, 22, 32, 50
Acaba, Joseph 152, 184, 195
Akers, Thomas 104, 108, 113, 126
Aldrin, Edwin "Buzz" 41, *51*, 60–1, *61*
Allen, Andrew 109, 120, 125
Altman, Scott 130, 138, 142, 153
Anders, William "Bill" 42, 56–7
Anderson, Clayton 149, 150, 155, 171, 172
Apollo 8, 10, 42–67, 70, 72
Apollo-Soyuz Test Project 72, 73, 77
Apt, Jerome "Jay" 105, 109, 120, 126
Armstrong, Neil 38, *51*, 60–1
Ashby, Jeffrey 136, 141, 144
Atlas rockets 20, 24

B

Baker, Ellen 103, 109, 123
Baker, Michael 107, 110, 122, 127
Barratt, Michael 156, 173, 174
Barry, Daniel 125, 136, 142
Bean, Alan 49, 62, *71*, 75
Blaha, John 103, 107, 113, 115, 126, 127
Bloomfield, Michael 129, 139, 143
Bluford, Guion "Guy" 88, 95, 106, 115
Bobko, Karol 87, 92, 95
Bolden, Charles 96, 104, 108, 120
Borisenko, Andrei 180, 182, 193
Borman, Frank 37, 56–7
Bowersox, Kenneth 109, 113, 124, 127, 167
Brand, Vance 77, 86, 89, 105
Brandenstein, Daniel 88, 93, 104, 108
Bresnik, Randolph 154, 194, 195
Brown, Curtis 109, 122, 126, 129, 131, 137
Buchli, James 92, 95, 103, 107
Budarin, Nikolai 118, 123, 167
Burbank, Daniel *135*, 138, 148, 183
Bursch, Daniel 112, 122, 126, 166

C

Cabana, Robert 104, 115, 121, 131
Caldwell Dyson, Tracy 149, 176
Cameron, Kenneth 105, 111, 124
Canadarm 85, 86, *101*, 120, 121, 140, 141, 143, *163*
Casper, John 110, 114, 120, 126
Cernan, Eugene "Gene" 39, *49*, 59, 67
Chamitoff, Greg 151, 152, 157, 172, 173
Chang-Díaz, Franklin 96, 103, 109, 120, 125, 131, 143
Chiao, Leroy 121, 125, 139, 160, 169; p 160
Chilton, Kevin 108, 120, 125
Clervoy, Jean-François 122, 128, 137
Clifford, Michael 115, 120, 125
Coats, Michael 90, 103, 106
Cockrell, Kenneth 111, 124, 127, 140, 143
Cold War 8, 12, 20, 24, 44, 77
Coleman, Catherine 124, 136, 177, 180, 182
Collins, Eileen 122, 128, 136, 146
Collins, Michael 33, 40, 46, *51*, 60–1
Collins, Sean 193, 194, 195
Conrad, Charles "Pete" *31*, 32, *33*, 36, 40, 49, 62, 74
Cooper, L. Gordon 21, 27, *31*, 32, 36
Covey, Richard 94, 102, 113, 115
Creighton, John 93, 107, 114
Crippen, Robert 83, 84, 87, 90, 91
Culbertson, Frank 112, 115, 165
Curbeam, Robert 129, 140, 148

D

Davis, Jan 109, 120, 129
Duffy, Brian 108, 112, 125, 139
Dunbar, Bonnie 95, 104, 109, 123, 130

F

Ferguson, Christopher 148, 152, 157
Feustel, Andrew 153, 157, 196
Fincke, Michael 157, 168, 173
Foale, Michael 108, 111, 122, 128, 129, 137, 168
Forrester, Patrick 142, 149, 154
Fossum, Michael 147, 151, 182, 183
Freedom 7 18, 22, 24
Friendship 7 21, 25

G

Gagarin, Yuri 10, 24, 80, 182
Garan, Ronald 151, 180, 182
Garneau, Marc 83, 91, 126, 139
Gemar, Charles "Sam" 107, 115, 120
Gemini 8, 14, 28–41, 44
Gernhardt, Michael 124, 128, 141
Gerst, Alexander 188, 189, 196, 197
Gibson, Robert "Hoot" 89, 96, 102, 109, 123
Gilruth, Robert 22, *33*, 61
Glenn, John 9, *21*, 22, *23*, 25, *33*, *119*, 131
Godwin, Linda 105, 120, 125, 142
Gorie, Dominic 131, 137, 142, 151
Grabe, Ronald 95, 103, 107, 112
Grissom, Virgil "Gus" *21*, *23*, 25, 30, 34, 52, *53*
Grunsfeld, John 123, 127, 137, 142, 153

H

Hadfield, Chris 124, 141, 185, 186
Halsell, James 121, 124, 128, 138
Harbaugh, Gregory 106, 110, 123, 127
Hartsfield, Henry "Hank" 86, 90, 95
Hauck, Frederick "Rick" 87, 91, 102
Hawley, Steven 90, 96, 104, 127, 136
Helms, Susan 110, 121, 126, 138, 164
Henricks, Terence 111, 115, 126
Hieb, Richard 106, 108, 121
Hilmers, David 95, 102, 107, 114
Hobaugh, Charles 141, 149, 154
Hoffman, Jeffrey 92, *101*, 105, 109, 113, 125
Horowitz, Scott 125, 127, 138, 142, *163*
Hubble Space Telescope 100, *101*, 104, 113, 127, 134, 137, 142, 153

I

International Space Station (ISS) 12, *13*, 17, 131, 134, *135*, 136, 138–99, *158*, 178
  2000–2010 158–77
  2011-2020 178–99
  Destiny module 140, 141, *163*
  Harmony node 150, 189
  Kibo (JEM) 151, *163*, 178, *181*, 184
  Leonardo module 142, 143, 147, 155
  Raffaello module 141, 142, 143, 157
  Rassvet module 156
  Rasvet module 176
  Tranquility node 155
  Unity module *13*, 131, *135*, 160
  Zarya module *13*, 131, *135*, 138, 160
  Zvezda module 138
Ivanishin, Anatoli 183, 192, 193
Ivins, Marsha 104, 109, 120, 127, 140

J

Jernigan, Tamara 106, 110, 123, 127, 136
Jett, Brent 125, 127, 139, 148
Johnson, Gregory 151, 153, 157
Jones, Thomas 120, 122, 127, 140

K

Kavandi, Janet 131, 137, 141
Kelly, Mark 142, 147, 151, 157
Kelly, Scott 137, 149, 177, *181*, 190, 191
Kennedy, John F. 8, 44, 46, 60
Kondratyev, Dmitri 177, 180, 182
Kononenko, Oleg 172, 183, 184, 190, 191, 197, 198
Kopra, Timothy 153, 191, 192
Korniyenko, Mikhail 176, 190, 191
Kotov, Oleg 171, 175, 176, 187

Kregel, Kevin 126, 130, 137
Krikalev, Sergei 118, 120, 131, 162, 164, 169

L

Lawrence, Wendy 123, 129, 131, 146
Lee, Mark 103, 109, 121, 127
Leestma, David *83*, 91, 108, 114
Lindsey, Steven 130, 131, 141, 147, 156
Linenger, Jerry 121, 127, 128
Linnehan, Richard 126, 130, 142, 151
Lopez-Alegria, Michael 124, 139, 144, 171
Lounge, John "Mike" 94, 102, 105
Lovell, James "Jim" *15*, 37, 41, 42, 56–7, 63
Low, David 104, 107, 112
Lu, Edward 128, 138, 167
Lucid, Shannon 93, 103, 107, 113, 125, 126
lunar module (LM) 46, 48, *51*
lunar rover 49, 65

M

Magnus, Sandra 144, 152, 157, 173
Malenchenko, Yuri 138, 167, 172, 184, 191, 192
Marshburn, Thomas 153, *163*, 185, 186
Mastracchio, Richard 138, 149, 155, 188
Mattingly, Thomas "Ken" 17, 66, 86, 92
McArthur, William 113, 124, 139, 170
McCall, Robert T. *15*, 67, 77, 84, 89, 156
McDivitt, James "Jim" 35, *51*, 57, 58
McMonagle, Donald 106, 110, 122
Meade, Carl 109, 115, 121
Melroy, Pamela 139, 144, 150, 172
Mercury 8, *9*, 18–27, 23, 44
Mercury Seven 20, 22, *23*, 131
Mir 116–19, 120, 122–31, 160, 165
Moon landings 8, 9, 44, 46, 60–2, *61*, 64–7
Mullane, Richard 90, 102, 114
Musgrave, Story 87, 94, *101*, 113, 115, 127

N

Nagel, Steven 93, 95, 105, 111
Nelson, George 90, 96, 102
Nespoli, Paolo 150, 177, 180, 182, 194, 195
Newman, James 112, 124, 131, *135*, 142
Nicollier, Claude 109, 113, 125, 137
Noguchi, Soichi 146, 175, 176
Novitskiy, Oleg 185, 193, 194
Nyberg, Karen 151, 186, 187

O

Ochoa, Ellen *101*, 111, 122, 136, 143
Oswald, Stephen 107, 111, 123
Ovchinin, Aleksey 192, 197, 198

P

Padalka, Gennady 168, 173, 174, 184, 190
Parazynski, Scott 122, 129, 131, 141, 150
Parmitano, Luca 186, 187, 198, 199
Pettit, Donald 152, 167, 183, 184
Phillips, John 141, 152, 169
Polansky, Mark 140, 148, 153
Precourt, Charles 111, 123, 128, 131
Progress 177, 183, 190, 191

R

Readdy, William 107, 112, 126
Reagan, Ronald 12, 80, 86, 100
Redstone rockets 20, *23*, 24–5, *45*
Reilly, James 130, 141, 149
Reisman, Garrett 151, 156, 172
Reiter, Thomas 147, 148, 170, 171
Richards, Richard 104, 109, 114, 121
Ride, Sally *81*, *83*, 87, 91
Robinson, Stephen 14, 129, 131, 146, 155
Romanenko, Roman 174, 175, 185, 186
Rominger, Kent 124, 127, 129, 136, 141
Ross, Jerry 96, 102, 105, 111, 124, 131, 143
Runco, Mario 110, 115, 126
Ryazansky, Sergey 187, 194, 195

S

Samokutyaev, Aleksandr 180, 182, 189
Saturn 11, *45*, 46, 47, 70
Schirra, Walter "Wally" *21*, 27, *33*, 37, 54, *55*
Scott, David 38, *51*, 58, 65
Searfoss, Richard 113, 125, 130
Seddon, Rhea 92, 106, 113
Sellers, Piers 144, 147, 156
Sharipov, Salizhan 130, 160, 169
Shepard, Alan *18*, *21*, 22, *23*, 24, *33*, 64, 182
Shepherd, William 102, 104, 110, 164
Shkaplerov, Anton 183, 189, 190, 195, 196
Shriver, Loren 92, 104, 109
Skripochka, Oleg 177, 192, 199
Skvortsov, Aleksandr 176, 188, 198, 199
Skylab 10, *11*, 68–76, *68*, 71, 73
Slayton, Donald "Deke" *21*, 52, 77
Smith, Steven 122, 127, 137, 143
Soyuz 12, 118, 160, 162
    MS 192–9
    TMA 164, 167–177, 181–192
Space Radar Laboratory 120, 122
space shuttle 10, *11*, 79
    Atlantis 95–6, *98*, 102–3, 105, 107–9, 114–15, *116*, 118, *119*, 122–9, *132*, 138, 140–1, 143–4, 148–50, 153–4, 156–7
    Challenger 78, 82, *83*, 87–91, 93–5, 97, 100, 102, 103
    classified missions 92, 95, 102, 114–15
    Columbia 10, 12, 14, 80, *81*, 82, 84–6, 88, 96, 104–6, 110–11, 113, 114, 120–1, 124–8, 130, 134, 136, 142, 145, 162
    Discovery 82, 90–3, 94, 100, *101*, 102–4, 106–7, 109, 111, 112, 115, 120–3, 127, 129, 131, 136–7, 139–40, 142, 146–8, 151–2, 154–6, 164–5
    Endeavour 11, 82, 108–10, 112–13, 120, 122–5, 126, 130–1, 137, 139, 141–4, 149, 151–3, 155, 157, 165–7, 182
    STS 13– *15*, 84–98, 101–115, 118, 120–131, 135, 136 138–144, 146, 148–159, 164–167, 172, 176, 182
SPACEHAB 112, 126, 127
Spacelab 88, 93–95, 109, 118, 126, 130, 134
Stafford, Thomas *33*, 37, 39, 59, 77
Stevens, Allen 52, 58, 59
Stott, Nicole 154, 156, 175
Sturckow, Frederick 131, 142, 149, 154
Sullivan, Kathryn *83*, 91, 104, 108
Surayev, Maksim 175, 188, 189
Swanson, Steven 149, 152, 188

T

Tanner, Joseph 122, 127, 139, 148
Thagard, Norman 87, 93, 103, 107, 118, 123
Thirsk, Robert 126, 174, 175
Thomas, Andrew 126, 130, 131, 140, 146
Thornton, Kathryn 108, 113, 115, 124
Thuot, Pierre 108, 114, 120
Truly, Richard 85, 88, 100
Tyurin, Mikhail 165, 171, 188

V

Virts, Terry 155, *181*, 189, 190
Volkov, Sergey 172, 182, 183, 191
Voss, James 115, 124, 138, *163*, 164
Voss, Janice 112, 122, 128, 137

W

Wakata, Koichi 125, 139, 152, 173, 174, 188
Walheim, Rex 143, 150, 157
Walker, Charles 90, 92, 96
Walker, David 91, 103, 115, 124
Walz, Carl 112, 121, 126, 166
Wetherbee, James 104, 110, 122, 129, 140, 144
Wheelock, Douglas 150, 176, 177
White, Edward "Ed" *9*, 35, 52, *53*
Whitson, Peggy 166, 172, 193, 194
Wilcutt, Terrence 122, 126, 130, 138
Williams, Jeffrey 138, 170, 175, 192
Williams, Sunita 148, 171, 184
Wilson, Stephanie 147, 150, 155
Wisoff, Peter 112, 122, 127, 139, 139
Wolf, David 113, 129, 130, 144, 153

Y

Young, John *9*, 10, *17*, *33*, 34, 40, 59, 66, 84, 88
Yurchikhin, Fyodor 144, 171, 176, 177, 186, 187, 194

ACKNOWLEDGMENTS

Whenever writers take on a work of nonfiction such as this, they stand squarely on the shoulders of earlier investigators and incur a good many intellectual debts. I would like to acknowledge the assistance of several individuals who aided in the preparation of this book. For their many contributions in completing this project I wish especially to thank recently deceased NASA Chief Archivist Jane Odom and her staff archivists at the NASA History Division who helped track down information and correct inconsistencies, as well as Bill Barry, Steve Garber, and Nadine Andreassen at NASA; the staffs of the NASA Headquarters Library and the Scientific and Technical Information Program who provided assistance in locating materials; Marilyn Graskowiak and her staff at the NASM Archives; and many archivists and scholars throughout NASA and other space organizations and archival repositories.

Assistance provided by a succession of interns also helped to make this book a reality: In addition to them, my colleagues at NASA and the Smithsonian Institution's National Air and Space Museum have always been helpful.

The collectors of mission patches are a continually active and helpful cadre of individuals. One of the leaders in this community is Robert Pearlman, creator of collectspace.com, who is a tireless advocate for the collecting community and a consistent help when seeking information about all things space-related. Gene Dorr, genedorr.com/patches/Intro.html, focuses on the intricacies of collecting mission patches for the "Golden Age" of spaceflight. Jacques Edwin van Oene and Erik van der Hoorn, spacepatches.nl/ have been collecting and designing space patches for decades. I thank them and many other collectors for their insights.

I wish also to thank the team at The Bright Press, especially Isheeta Mustafi, Sarah Herman, Jacqui Sayers, Caroline Elliker, and Polly Goodman, who improved this work and made it possible to bring it to fruition.

Special thanks to A-B Emblem, especially John Lindsey, for assistance in securing images for several of the mission patches appearing in this book.

DR. ROGER D. LAUNIUS, a U.S. government historian for 35 years, has served as chief NASA historian and worked at the Smithsonian Institution's National Air and Space Museum, most recently as Associate Director for Collections and Curatorial Affairs. He is a Fellow of the American Association for the Advancement of Science, the International Academy of Astronautics, and the American Astronautical Society; and Associate Fellow of the American Institute for Aeronautics and Astronautics. He has written and edited over thirty books on aerospace history and has been a guest commentator on National Public Radio and major television news networks.

KATHRYN SULLIVAN, a geologist and former NASA astronaut, became the first woman to walk in space in 1984. She has since served as founding director of the Battelle Center for Science, Engineering and Public Policy at the Ohio State University, the Under Secretary of Commerce for Oceans and Atmosphere and Administrator of the National Oceanic and Atmospheric Administration, the Lindbergh Chair in Aerospace History at the Smithsonian Institution's National Air and Space Museum, and Senior Fellow at the Potomac Institute for Policy Studies.

GEMINI VII

GEMINI IX

EXPEDITION 14

EXPEDITION 16

EXPEDITION 17

EXPEDITION 18

EXPEDITION 19

EXPEDITION 20

EXPEDITION 35

EXPEDITION 60